INFORMATION
DOESN'T WANT
TO BE FREE

Information Doesn't Want to Be Free

Laws for the Internet Age

Cory Doctorow

McSWEENEY'S

SAN FRANCISCO

McSWEENEY'S

mcsweeneys.net

Copyright © 2015 Cory Doctorow

Cover design by Sunra Thompson.

McSweeney's and colophon are registered trademarks of McSweeney's,
an independent publisher with wildly fluctuating resources.

Printed in the United States

ISBN 978-1-940450-46-9

10 9 8 7 6 5 4 3 2 1

Neil Gaiman

George shook his head slowly. "You're wrong, John. Not back to where we were. This morning, we had an economy of scarcity. Tonight, we have an economy of abundance. This morning, we had a money economy—it was a money economy, even if credit was important. Tonight, it's a credit economy, one hundred percent. This morning, you and the lieutenant were selling standardization. Tonight, it's diversity. The whole framework of our society is flipped upside down." He frowned uncertainly.

"And yet, you're right, too. It doesn't seem to make much difference. It's still the same old rat race. I don't understand it."

—from "Business as Usual, During Alterations,"
by Ralph Williams (*Astounding Science Fiction*, 1958)

I BOUGHT A box of SF pulps when I was in my late teens from one of my father's friends, who kept them in the garage. English editions of *Astounding Science Fiction*, for the most part. Stories written by authors whose names I barely recognized, despite being a science-fiction reader from about as soon as I could read.

I paid more than I could afford for them.

I suspect that one story paid for all of them, though.

It's a thought experiment. Until recently I'd forgotten the opening of the story (aliens decide to Mess With Us) but remembered what happened after that.

We're in a department store. And someone drops off two matter duplicators. They have pans. You put something in pan one, press a button, and its exact duplicate appears in pan two.

We spend a day in the department store as they sell everything they have as cheaply as possible, duplicating things with the matter duplicator, making what they can on each sale, and using checks and credit cards, not cash (you can now perfectly duplicate cash—which obviously is no longer legal tender). Toward the end they stop and take stock of the new world waiting for them, and realize that all the rules have changed, but that craftsmen and engineers are more necessary than ever.

They realize that companies won't be manufacturing millions of identical things, but will need to make hundreds, perhaps thousands, of slightly different things. That their stores will be showrooms for things, and stockrooms will be history. That there will now be fundamental changes in 1950s-style retailing—including, to use a phrase that turned up well after 1958, a long tail.

Being *Astounding Science Fiction*, the story contains the moral of 95 percent of *Astounding Science Fiction* stories, which could perhaps be reduced to: People are smart. We'll cope.

When my friends who were musicians first started complaining sadly about people stealing their music on Napster, back in the 1990s, I told them about the story of the duplicator machines. (I could not remember the name of the story or the author. It was not until I agreed to write this foreword that I asked a friend, via email, and found myself, a Google later, re-reading it for the first time in decades.)

It seemed to me that copying music was not stealing. It was something else. It was the duplicator machine story: you were pressing a button and an object appeared in the pan. Which meant, I suspected, that music-as-object (CD, vinyl, cassette tape) was going to lose value, and that other things—mostly things that could not be reproduced, things like live shows and personal contact—would increase in value.

I remembered what Charles Dickens did, a hundred and fifty years before, when copyright laws meant that his copyrights were worth nothing in the U.S.: he was widely read, but he was not making any money from it. So he took the piracy as advertising, and toured the U.S. in theaters, reading from his books. He made money, and he saw America.

So I started doing Evenings with Neil Gaiman as fund-raisers for the Comic Book Legal Defense Fund, and learning how to do that—how to make an evening interesting for an audience, with just me and a stage and things I'd written, partly because it seemed to me that one day it might not be as easy to make money from selling stories in the traditional way, but that business might still continue more or less as usual, during the alterations, if there were other things I could do.

And so as the nature of music-selling changed utterly and fundamentally, I just stood and watched and nodded. Now the nature of book publishing is changing, and the only people who claim to know what the landscape of publishing will look like a decade from now are either fools or deluding themselves. Some people think the sky is falling, and I do not entirely blame them.

I never worried that the world was ending, because as a teen I'd read a thought experiment in an SF pulp published two years before I was born. It stretched my head.

I know that the view is going to be very different in the future, that authors are going to get their money from different places. I am certain that not all authors can be Charles Dickens, and that many of us became authors in order to avoid getting up on stages in the first place, and that it's not a solution for everybody or even for most of us.

Fortunately, Cory Doctorow has written this book. It's filled with wisdom and with thought experiments and with things that will mess with your mind. Once, while we were arguing, Cory came up with an analogy that explained the world we were heading into in terms of mammals and dandelions, and I've never seen anything quite the same way since.

Mammals, he said, and I paraphrase here and do not put it as well as Cory did, invest a great deal of time and energy in their young, in the pregnancy, in raising them. Dandelions just let their seeds go to the wind, and do not mourn the seeds that do not make it. Until now, creating intellectual content for payment has been a mammalian idea. Now it's time for creators to accept that we are becoming dandelions.

The world is not ending. Not if, as *Astounding Science Fiction* used to suggest, humans are bright enough to think our way out of the problems we think ourselves into.

I suspect that the next generation to come along will puzzle over our agonies, much as I puzzled over the death of the Victorian music halls as a child, and much as I felt sorry for the performers who had needed only thirteen minutes of material in their whole life, and who did their thirteen minutes in town after town until the day that television came along and killed it all.

In the meanwhile, it's business as usual, during alterations.

Amanda Palmer

I'M A STREET performer. Nowadays I do other things too. I'm a rock star as well, and I run the business that is my life and my job. But before I was a rock star, between about 1998 and 2002, street performing was how I paid my rent, bought my beer, and fed myself. My act was called "The Eight-Foot Bride." I was a Living Statue—you've probably seen something like it. I stood on a box, white from head to toe, dressed as a mournful bride in a Victorian wedding gown and bridal veil (purchased for $19.95 at a Boston thrift shop), accessorized with long white gloves ($9.95, purchased at the fabric store in Waltham, replaced every forty performances or so) and a black wig ($29.95, purchased at Dorothy's Boutique, in Boston, replaced about twice a year). I came to life and sorrowfully handed a white daisy-pom flower (purchased at discount for $2 per bunch from a sympathetic florist on the corner who gave me his day-old rejects) to each and every person who dropped any amount of money into an old-fashioned tin milk jug (stolen from my folks' house) that lay at my feet.

I would stand there for hours at a time, watching the habits of passersby.

Of the literally *millions* of people who walked by my humble performance in dozens of cities on three different continents:

- Some didn't look at me, and gave me nothing.
- Some looked at me, enjoyed my performance, and gave me nothing.
- Some looked, enjoyed, and gave me a small amount of money.
- Some looked, enjoyed, and gave me a huge amount of money. (I was tipped with a twenty-dollar bill about every four or five performances. This was always a cause for celebration.)

- Some looked, enjoyed, and didn't have any money, but left me thank-you notes, sketches, poems… or got more creative and left pumpkins, fountain pens, cigarettes, or, in one very dubious act, a pink, powdery dime bag of unidentified drugs (the effects of which are a story for another foreword).

Here was the shocker (at least to those with the general music-industry mentality of the 1990s):

I made a steady, predictable income every single day.

I could pretty much count on making twenty to thirty dollars an hour on an average weeknight in Harvard Square. And more like forty to fifty dollars an hour on the weekend nights.

Like clockwork, people were generous. Nobody asked them to be. I just stood there, literally silent, waiting for them to tip me out for the weird, loving act of randomness I was making available to humankind. Many times some stranger would sit and watch me from a distance for five or ten minutes. In those five or ten minutes, they became fans. They would then leave me a dollar (or more, or less) and go their own way, never to return.

I maintain that no job could give today's musicians a better education than four years of street performing. Nothing could reeducate the crumbling and scrambling music industry more than a few years of hawking their talent to the wide world of international passersby. This is what they would learn, and this is what they could apply to the Internet, and this is what Cory Doctorow understands after years in the world of copyfighting:

People actually like supporting the artists whose work they like. It makes them feel happy. You don't have to force them. And if you force them, they don't feel as good.

This is fundamental.

Back when I was in my first band, the Dresden Dolls, we were signed to a major label. It was 2004, our debut CD was popular and selling in stores, but it was also the beginning of the era of home CD-burning and torrenting.

And not once, not twice, but dozens of times, I had fans come up to me in signing lines and hand me cash in ten- and twenty-dollar bills (once someone even wrote me a personal check and demanded I keep it), insisting that they wanted to absolve themselves of the guilt of having burned our CD. Many times, people said, "I tried to find your music in a store, but I couldn't. I really wanted to be able to buy it. Please, please, take this money."

I never suggested they do this. They came to me. They'd enjoyed the content, and they wanted to feel the pleasure of firsthand, direct support for the content provider.

Neil told me Cory's metaphor about dandelions and mammals a few nights ago. I couldn't help but think that my time standing still in the street wasn't all that different, in a kind of poetic way—except I was the empty dandelion stalk, standing in silence, watching the sea of humanity pass, holding on to a faith that a few seeds out of thousands would blow back in my direction. I knew many wouldn't. But I needed only about 3 percent of the passersby to take part in my magic little performance to pay my rent and feed myself. And they always did.

As long as people make art and content, and other people want art and content, the marketplace will adjust to create paths for them to connect and support each other.

For anyone who has been claiming that a more "faith-based" honor system on the net is too theoretical and utopian ("Sure, it would be *nice* if people actively kicked back money to digital content creators when they didn't *have* to, but they *won't*... that's wishful thinking"), I point them back to the philosophy that has kept street performance viable for hundreds of years. Not everybody passing by will play the game. But enough people will play to make free content an ongoing reality.

Those people aren't pretend, theoretical people. They exist here, now, on the planet, surfing the Internet, absorbing content, information, and art. They actively, vocally offer to give support to the many entities online—artistic and otherwise—that keep the Internet, and the world, a generally nicer and more interesting place to be.

The minute you start locking up the pathways of the Internet, and putting rules and regulations and fences around content that was once freely sharable, you ruin the possibility of an entire generation organically evolving into a more mutually supportive ecosystem. We don't know exactly what it will look like or how it will function, but as long as human beings enjoy trafficking in the commerce of art and information, it is possible.

In street terms: locking up Internet content is the equivalent of issuing mandatory ball gags for street musicians, to be worn until an interested passerby brokers a deal with a middleman who has a permit to remove the ball gag and release the musician for a few minutes, long enough to let them play a song, after which point their hands are tied up again and the ball gag goes back into the mouth. By putting laws into place that stop the free flow of information, sounds, and images, you mute the possibility of a real, authentic exchange. You take control *out* of the hands of the content creator, and *out* of the hands of the public, who can no longer decide for themselves how they want to offer their support.

Two years ago, I conducted a crowdfunding campaign that raised over a million dollars in capital so that I could put out a record without a major label. People scratched their heads—why would this happen? How did she do it? The newspapers, the journalists, the bloggerati all weighed in. Was this the future of music? Was my Kickstarter "repeatable"? Am I a freak, an outlier, a strange charity case that an outlying public accidentally raised above the norm?

Not from where I'm standing. There are many more of me—there already have been, and we are legion. It's repeating as we speak.

We are a new generation of artists, makers, supporters, and consumers who believe that the old system through which we exchanged content and money is dead.

Not dying: dead. Not savable, not reinstatable, not resuscitatable. And this fact doesn't cause us agony. (Oh no! What the fuck are we going to do without labels to truck our music around and send us checks?) We see it as a chance to celebrate our freedom. (Fuck yes! We can give our art away, for free, to whomever we want! And they can put money directly into our pockets! Without someone telling us how to make art or how to pay for it!)

Of course, no artist's approach is exactly repeatable, just like no fanbase or community will ever act or react the same way twice.

Will there ever be another Grateful Dead? Will there ever be an exact duplicate of the Deadheads?

No way. It's a onetime deal. As it is with any great content creator, or any large community. But there are patterns, and we learn lessons.

This is the lesson I learned, as a rock star and as a street performer in my wedding dress:

- Keep the content authentic,
- keep the exchange honest,
- keep the message spreading by any means necessary,
- and people will come.
- Once they come, if you make it easy for them, many will pay.

When people feel and know that you are keeping the channels open, the doors open, the airwaves unblocked, the locks unlocked... they come. And they will pay their hard-earned cash to keep the cycle continuing.

Just as I could count on a predictable income in the street, it's not a stretch to say that humankind will adjust to freely available content, and react to the demands of its content-makers as needed.

The musicians I see trying to keep content locked up are generally the ones who aren't creating any new content—and who aren't hanging out on the Internet very much, where they could see the enthusiasm and goodwill of their fans firsthand. They don't see the open windows, only the closed doors. They're trying to keep things as they are because they don't know any other system. They've been getting their royalty checks from labels for CD and vinyl sales for years, and the numbers are dropping. They face an uncertain future if they can't count on those checks.

So they argue and they fight and they try to make it 1990 again. But 1990 isn't going to come back. And most of the younger musicians I know nowadays don't even realize there's anything to argue about.

They're too busy making music and uploading it, blasting it out to their small networks, and giving it away for free in order to convince people to become a part of their story, their tribe.

They take it for granted that if the content is worthwhile, their careers will take off, and they'll somehow make a living.

And they're right.

Trying to protect a system that's now fundamentally broken is like trying to reroute a raincloud to go and thunderstorm over a different town. You're better off dealing with the facts, and grabbing your umbrella.

Or, if you're like many of the people I know: stripping naked and running around in the street, screaming with joy, and enjoying the downpour.

COPY! COPY! COPY! COPY! COPY! COPY!

Detente

THE COPYRIGHT WARS are nothing new. Five hundred years ago, Europe was convulsed by a struggle to determine who could access the Bible and under what circumstances. Canterbury and Rome spent decades battling over whether the uncertain benefits of broad access to scripture were worth diminishing the undeniable majesty of their incumbent religious institutions. A few hundred years later, at the turn of the twentieth century, composers were forced to defend their own turf—they called performers pirates, and insisted that recording music was a form of theft. John Philip Sousa, the great American composer, thought the record player would make musicians extinct: "Today you hear these infernal machines going night and day," he said in 1906, while testifying before Congress. "We will not have a vocal cord left. The vocal cords will be eliminated by a process of evolution, as was the tail of man when he came from the ape."

A few decades after that, the record men were struggling with their own challengers. In the 1930s, ASCAP, the organization that licensed "mainstream" recording artists' broadcast rights to radio, raised its fees by nearly 500 percent. The radio stations responded by boycotting ASCAP. That meant that ASCAP artists didn't get played on the radio. Instead, within a few years, the stations turned to another, newer artists' organization, BMI, which had been created to represent "race artists" and "hillbilly music"; for much of 1941, the genres that ASCAP neglected were the only ones that got airtime, and so only those upstarts saw a dime.

After ten months of this, ASCAP's artists rebelled. Why should these other musicians—whose legitimacy ASCAP didn't even want to acknowledge—get a never-ending payday while they starved? ASCAP backed down on its fee increases, and the most important principle in the copyright wars was affirmed: Money talks and bullshit walks.

If you want to attain fame and fortune as a creative person, you're probably like I was when I started out: a harried, busy individual trying to squeeze in time to make art around the constraints of a day job and family. You don't want to hear about the copyright wars. You certainly don't want to fight in them. You want to *make stuff*, and, ideally, get paid for it.

I'm with you.

This is a book about the reality, today, of the Internet and the regulations that surround it, and the ways that those regulations shape successful strategies for earning a creative wage. It is also a book about the profound pitfalls that both creators—writers, musicians, filmmakers, painters—and their investors—publishers, studios, record labels—fall prey to, when it comes to getting their work out into the world.

It's been more than five hundred years since the copyright wars began, and they show no sign of abating. Peace in our time seems unlikely. If you want to make stuff and try to earn a living from it, rather than shaking your fist and telling the Internet to get off your lawn, then this is the book for you.

Creators, investors, intermediaries, and audiences

When we're talking about copyright, we're fundamentally talking about four different activities: making creative works, investing in creative works, distributing and selling creative works, and enjoying creative works. As a shorthand, I'll be using "creator" to describe people who make creative works—painters, photographers, game designers, novelists, poets, musicians, songwriters, choreographers, dancers, and many other sorts of people. I'll use "investor" to describe someone who puts capital—cash—into the production and refinement of that work: think of a publisher, a record label, a studio. I'll use "intermediary" to describe those entities that handle the work between creation, investment, and delivery: a distributor, a website like YouTube, a retailer, an e-commerce site like Amazon, a cinema owner, a cable operator, a TV station or network. Finally, I'll use "audience" to describe the person the work ends up with. These are all fluid roles, of course: some intermediaries (like cable operators) are also investors or creators; some creators (like film directors) require investors before they can make anything; and some of the most dedicated audiences are made up of creators themselves (which is why authors are usually photographed standing in front of a wall of other people's books).

What Makes Money?

THERE IS NO real secret to success in business. If you want to make as much money as possible, all you need to do is make something people are willing to pay for, and offer it at a price they're willing to pay. Complaining about the universe's unfairness is never part of a successful strategy.

Here are some other things that don't make money:

- Complaining about piracy.
- Calling your customers thieves.
- Treating your customers like thieves.

A common refrain about entertainment and business in the news today is how much money all those investors and creators *could* be making if only the Internet and its users would start behaving themselves. If you've already made a cool couple billion in the old world, have plenty left in the bank, and want to take a gamble on getting your buddies in Congress to pass laws to suit your needs, this kind of complaining might seem like a winning strategy.

But if you want to start earning a creative living *now*, you can't play by those rules. Wouldn't it be nice to come up with a way of

Money on the table

When a creator tries to make money, she becomes an entrepreneur—a businessperson. Businesspeople are prone to all sorts of madness. Succeeding in business requires that you avoid this madness. One particular strain of madness is the overwhelming, irrational concern that you might be letting someone benefit from your work for free—what an economist would call "aversion to positive externalities." In plain English, that's worrying that someone else is getting some benefit from your investment of labor or capital. It's a bit like worrying that lost strangers are reading their street maps using the extra light from your porch, as you sit out reading on a warm night—if you're paying for the light, why should they reap the benefit? If you can figure out a way to set up a service selling porch light, you might be onto something. But if you find yourself resorting to angrily switching off the light every time someone gets too close to your house, you're not going to have a very pleasant night. When it comes to policing your externalities, a little goes a long way.

making money that works on the Internet we have *today*, instead of the one that Hollywood is betting on creating *tomorrow*?

Jam tomorrow

My grandmother grew up in Leningrad, in the former Soviet Union, and is a wellspring of dour sayings from the old country. One of my favorites is "Jam yesterday; jam tomorrow; no jam today." That is, we used to have jam, before everything changed, and we're told there will be even more jam in the future. But right now, there is no jam. Or: Don't tell me about pie in the sky when I die. I'm hungry *now*.

Don't Quit Your Day Job—Really

BEING AN ENTREPRENEUR is a gamble. 66 percent of all new businesses fail in the first four years, and that's across *all* sectors, from import/export to running a restaurant to starting a babysitting service or an e-commerce site. Creative businesses are even more risky, because the reasons we make art are fundamentally different from the reasons we undertake other kinds of commercial activity. You might, from child-hood, nurse a dream of being a restaurateur, but if you looked around and found that your city was full of failing restaurants whose owners were going bust, you'd probably pick another career.

In the creative arts, though, the lack of any reasonable market-place for one's "product" is rarely a good enough reason to give up on it. Look at poetry (to name just one field): there was never much of a market for poetry, and today the market is as bad as ever. There's prac-tically no commercial poetry. But today, more poetry is being written than ever before. People spend real, folding money to learn to write better poetry. They devote hours and days and weeks of their lives to it. They are driven by it. In many cases, they can't stop.

Humans seem to make art reflexively. Small children sing songs, make drawings, and tell stories long before they're exposed to the idea of money, markets, or remuneration. People working through trauma and depression have been rehabilitated by art therapists teaching them to express themselves artistically. There is something inherently *non-market* about making art. It's something that economists would call "irrational," in that the work artists put into their craft exceeds any reasonable expectation of profit or even a break-even return.

I put my first short story in the mail when I was sixteen. I sold my first short story a year later, to a "little" market (one that didn't pay professional rates). *Ten years later*, I made my first sale to a market that qualified as "professional" by the lights of the Science Fiction Writers of America, paying something like four cents a word. *Five years after that*, my first novel came out, for which I received a little more than seven thousand dollars. I was thirty-one, and fifteen years had gone by between my first attempt to be a "professional" writer and my first published novel. Five years after that, a full twenty years after I started trying to get paid to write, I quit my day job.

But through it all, I couldn't stop. Seriously: if you'd broken all my fingers, I would have dictated my stories, or written them with a pen clenched between my teeth. This doesn't mean it's always fun— my family knows that I can be a little irritable when I'm on a difficult deadline. I wrote a 120,000-word novel in the last three months of 2011, and it made me really grumpy, especially toward the end. Merry Christmas, now leave me alone, I'm on a deadline.

But whatever personal flaws I exhibit when I'm working are as nothing compared with the absolute, remorseless orneriness on exhibition when I'm *not* working. My interior world consists, largely, of thinking of books to write, writing books, recuperating from writing books, and then thinking of more books to write. In between, I write essays, speeches, novellas, novelettes, short stories, investigative journalism, columns, and blog posts. Lots of blog posts. I'm the coeditor and co-owner of a website called *Boing Boing*, and I've spent the past ten years writing about ten blog posts a day, every day. I often find myself unable to think about anything in any depth without writing a blog post about it to check whether I understand it sufficiently to convey it to someone else.

I am *compelled* to write. Long before I wrote to keep myself fed and sheltered, I was writing to keep myself sane.

Whatever kind of arts career you're hoping for, the odds are against you. This is a serious downer. Pretty much everyone who ever set out to earn a living (or part of a living) in the arts failed. "Don't quit your day job" isn't just a sarcastic barb to toss at your friends when they play their guitars at you: it's goddamned *great* advice, especially if you're hoping to support a family or save for your old age.

When I set out to be a "professional writer" (filled with naive and downright weird ideas about what this was going to entail), I knew, intellectually, that I was probably not going to make it. On the other hand, I also knew that I was perfectly miserable when I wasn't writing, and that I yearned for widespread publication with all my heart.

If you are like that, if you can't stop, and you are planning to try to earn a living off your art, then I want you to know some things about the Internet. Because the marketplace for creative works today involves the Internet, in ways large and small. You might be handcrafting one-of-a-kind sculptural-assemblage birdhouses out of rusty license plates in your Unabomber shack in the middle of the woods, but chances are that you sell them through a site like Etsy. You might be recording a collection of all-acoustic roots music using instruments that predate the invention of piano rolls, but chances are you're going to start your research with Google, find your collaborators with Craigslist, and sell your music as MP3s.

If you're setting out to earn a lot of money in the arts, your chances are slim. There are *some* artists who "make it," for whatever definition of "make it" you favor. A tiny number become rich. A small number earn a good living. A slightly larger number earn a crappy living. And a respectable number earn *something*. But if you're setting out to earn a lot of money in the arts without regard to how Internet regulation shapes the marketplace for creativity, your chances are *zero*.

The single most important thing you need in order to have a career in the arts is persistence. The second most important thing you need is talent. The *third* most important thing is a grounding in how the the online world works. It's that important.

The goal of this book is to provide a first-of-its-kind look at the pitfalls and opportunities for earning a creative living in the age of the Internet. I want to equip you with the critical skills required to have a non-zero chance of making a living as an artist today, in the world as it is. Not in the world as it was in the pre-Internet era, and not in any of the tomorrows we've been promised.

What I do on the Internet (aka: Why listen to me?)

Why should you listen to what I have to say about the Internet? Well, I may not be the world's geekiest artist (I hold out novelist Charles Stross or cartoonist Randall Munroe for this honor), but I am a pretty serious geek. I dropped out of university to be a computer programmer, cofounded a software company during the first dot-com boom, sold it, went to work for the Electronic Frontier Foundation (a civil rights organization that works on tech issues), helped build one of the most successful author-owned websites in the world, and pioneered electronic fiction distribution.

I've been on the *New York Times* best-seller list twice; I've also released my novels and short-story collections as free downloads simultaneously with their print editions. My last two novels have drawn mid-six-figure advances. I've won a long list of awards, and been nominated for a longer list.

I hold an honorary PhD in computer science, and *Boing Boing*, the website I co-own, grosses several million dollars per year. I was a delegate to the United Nations World Intellectual Property Organization, I write columns and lecture pretty much everywhere, and every now and again I teach writers, prospective technologists, and even lawyers. If I'm wrong about how the Internet works, it's not due to lack of opportunity to learn, or for having failed to think about it. If I'm wrong about how an artist's life works, it's not for lack of making a living as one. And if I'm wrong about how copyright law interacts with the business of creativity, it's not for lack of experience with both.

In short: if I'm wrong, I promise that I'll be wrong in a well-informed and interesting way.

Any Time Someone Puts a Lock on Something That Belongs to You and Won't Give You the Key, That Lock Isn't There for Your Benefit

I gave a speech to a bunch of publishers at the O'Reilly Tools of Change for Publishing Conference in New York in 2009, and halfway through, I said: "I'm not often certain about something, but I'm certain of *this*. I'm so certain of it that I'm willing to call it 'Doctorow's Law.'" Everyone laughed. Later, I mentioned this casually to my agent, Russ Galen. (I like to repeat my laugh lines until they're worn rather thin, I'm afraid; it's a habit I inherited from my grandfather, who liked to grind away his jokes to micron-thicknesses.) Russ sat up very straight and said, "No. You must have *three* laws." Russ represented Arthur C. Clarke through much of his career, and continues to represent Clarke's estate, and of course Clarke got a lot of mileage out of "Clarke's Three Laws," so I took the advice to heart.

E VERY GRAND THEORY has to start somewhere, ascribe some first cause to all that comes after. It's a bit of an artificial exercise, of course, because everything complicated has complicated causes. But you need to start somewhere.

I'm going to start with the World Intellectual Property Organization. Founded in 1967 as a successor to the Bureaux Internationaux Réunis pour la Protection de la Propriété Intellectuelle (the BIRPI, then just shy of its seventy-fifth birthday), WIPO was later subsumed into the United Nations as a "specialized agency." WIPO writes the world's major copyright treaties (with two recent and important exceptions, ACTA and the TPP, that I'll get to later), and in that regard it has the same relationship to stupid copyright laws that Mordor has to evil in Middle-earth. It is the origin, the prime mover.

In 1996, WIPO approved the WIPO Copyright Treaty, or the WCT, and its cousin, the WIPO Performers and Phonograms Treaty (WPPT). Collectively, they are called "the Internet treaties" by international copyright wonks, and if you want to understand the market for creative works on the Internet—the market for movies, books, songs, and anything else that can be sold digitally—you have to understand something about these treaties.

Specifically, you need to understand one key aspect of the WCT: anti-circumvention. That's what I'm going to discuss in this chapter.

The WCT is everywhere

When the UN writes a treaty, it asks the countries of the world to sign on to it. These signatories go on to pass national laws that reflect their obligations under these treaties. In the USA, the WCT was enacted into law in 1998 as the Digital Millennium Copyright Act (the DMCA). A few years later, the European Union created the EU Copyright Directive (EUCD), which each EU country then turned into law in *its* national body. What that means is that most of the world's industrialized countries have some version of the WCT on the books. These implementations of the WCT vary, but usually the variations are small. For that reason, I'm going to talk about the effect of the WCT as a global phenomenon, not limited to the USA, the EU, Australia, Japan, or anywhere else.

Anti-Circumvention Explained

WHAT THE HELL is anti-circumvention? Loosely speaking, it's a law against "circumventing" (getting around) a digital lock. Pretty much everyone in the world has experienced a digital lock. Ever tried to fast-forward through the anti-piracy warning at the start of a DVD and gotten an ACTION NOT ALLOWED message? That's a digital lock. In order to legally allow you to descramble the movie on your DVD, your DVD-player manufacturer has to sign a licensing agreement that says, "I promise I will design my player so that users can't skip anti-piracy warnings."

It's not hard to make a DVD player that ignores this requirement. DVD encryption was first broken fifteen years ago, and there are now tons of software-based DVD players (like HandBrake) that can play back any part of a DVD at any speed, no matter what "region" it's from. (DVDs contain a region code meant to prevent discs from being bought in one part of the world and played in another.) These players also let you save DVDs to your hard drive, copy them to your tablet or phone, and back them up to your home media server.

Curiously, there's no copyright law that says it's illegal to skip piracy warnings. There *is* a law (the DMCA, in the U.S.) that says it's illegal to descramble a movie without permission—and this is what makes anti-circumvention so pernicious. Fundamentally, anti-circumvention is a way of making up new copyright laws. You can prohibit quotation, eliminate the rewind button, limit pausing to ten minutes, and force viewers to sit through ads, just by scrambling your art and then saying to the player manufacturers, "You can't descramble this unless you honor my requirements."

This may sound like a pretty good deal for creators and their investors: after all, it lets you charge money for stuff that the audience used to get for free. When you sold someone a book, back in the day, they got the whole thing in one package. With digital locks, you can sell them only the right to look at the book after 6 p.m., while physically located in North America and not in a commercial establishment. If they want the "read on the subway" rights, those can be sold separately.

I'll come back to what this means for audience members later in this book, but for now let's stick with the impact this has on creators and investors. Is it really a good deal for them? Or has anti-circumvention ended up taking negotiating power away from people who make and finance art, and handing it to companies that make digital locks?

Crypto 101

You don't have to be a cryptographer to understand this stuff, but you do need to have a conceptual grasp of the way that scrambling and descrambling work. Fundamentally, this is about four things. First, there's the unscrambled file, called the "cleartext," which would be the movie you watch, the song you hear, or the book you read. Next, there's an "algorithm"—a mathematical system for scrambling the text. These algorithms are usually public and well understood, because no one is ever sure if an algorithm is secure until all her peers get a chance to look at it and check it over for flaws. Third, there's the "key," a secret password that's fed to the algorithm along with the cleartext in order to produce the scrambled file. This scrambled file (our

fourth thing) is called the "ciphertext," and in that state it's an indecipherable mess of no value to anyone unless they have the key to descramble it. People can copy this encrypted file all they want, and it won't ever substitute for the legit product, because it's indistinguishable from random noise. If the algorithm works, the file can't be unscrambled unless you have that key.

With a digital lock, the player—a Kindle, a DVD player, a software program that plays or displays locked files—has a copy of the secret key embedded in it. You load the ciphertext file onto the player, it unscrambles the file and lets you see the cleartext, and then it throws the cleartext away when you're done.

How compatibility works

If you've been using computers for a long time, you probably remember when there were multiple, incompatible ways of creating documents. You'd

send someone your AppleWorks spreadsheet, and they'd complain that they only had Excel. You'd send them a Word document, and they'd only have

WordPerfect. Over time, this ceased to be a problem, mostly because these programs started including an automatic feature for reading and writing each other's files. At first, this was complicated: you'd try to open a word-processor document, and your program would ask you if you wanted it converted from some obscure file format into one your program could understand. Then it became automatic: your computer stopped asking and just automatically opened every file you fed it, converting as need be.

Likewise, in the early days of web browsers, there were lots of warring graphics formats, a whole acronym-salad's worth: GIF, JPEG, PNG, SVG, BMP. Depending on the page you visited and the browser you used, you'd get little red x's or error messages telling you your browser wasn't compatible with someone's animated dancing hamster image. But today your browser just reads and displays all those files without batting an eyelash.

What happened? Well, programmers made their programs compatible with the other formats. This is something that is very hard to do in the physical world (making your toaster compatible with your dishwasher isn't a project for the faint of heart), but it's routine in the digital world.

But not where digital locks are concerned. Here's where the problem for creators and investors comes in. Because anti-circumvention rules mean that only a digital lock's maker can authorize you to open it, on-the-fly conversions to improve compatibility often aren't allowed. That's how copyright ceases to protect creative works and begins to protect digital locks instead.

Is This Copyright Protection?

THE PEOPLE WHO make digital locks sell them as "copy protection" (that is, protection against having a file copied), and sometimes as "copyright protection." We can debate their claim to the former, but we should certainly reject the idea that digital locks protect copyright. As things stand now, it's the other way around.

Many different reasons and rationales for copyright have been offered since its inception. England's Statute of Anne (1710) set out to protect the established English publishers from Scottish competition. Sixty-some years later, the U.S. Constitution provided for copyright "to promote the Progress of Science and useful Arts." The Berne Convention for the Protection of Artistic and Literary Works, one of the first international copyright agreements (its text was based on a draft created by Victor Hugo in 1878), added protection of an author's "moral rights"— the right to claim authorship of a work, for example, and to prevent its distortion or modification.

One rationale that has *never* been offered is that copyright exists to protect middlemen, retailers, and distributors from being out-negotiated by creators and their investors.

Now, let's say there's a company with a wildly popular video-distribution technology. If you're a filmmaker, they'll sell your movies and give you 70 percent of the revenue, and even better, they'll promise you that they'll keep your movies safe from piracy by putting their digital locks on them. At first, everything's great. You're making money, they're making money. You sell a million movies. Five million. Ten million!

But then they put the squeeze on: the 30-70 split was just an introductory offer. Now that they've proven how great their technology is, they want a 50-50 split. At the current rate of sale, that means you'll be giving them millions of dollars every month: money you were putting back into your own business, using to create more art, using to pay your own overheads.

Without anti-circumvention, this is easy: you just call a meeting with the video distributor and say, "No deal. First of all, there're plenty of other operators out there who can do what you do. You're the distributor—your job is to provide invisible, commodity plumbing that puts our movies in front of our audience. Second of all, for the money you're talking about, we could just skip the middleman and build our *own* system. So yeah: no deal." It's your copyright, after all; you have all the negotiating leverage.

But this works only if your audience can follow you from the old format to the new one. If I'm your audience, and I've spent a thousand bucks on my movie library, I don't want to have to throw away that investment. I want to be able to use one family of devices and one program to manage my movies. I'm not going to try to remember which movie goes with which player and which device. And I'm certainly not going to get a new box under my TV just for your movies (let alone a new TV!).

This is how, once you add anti-circumvention to the mix, all of copyright's protection is handed directly to the company that slapped a digital lock on the product in question. The filmmaker can't authorize audiences to break those locks and convert their movies to play on a new device. The investor can't authorize that either. Only the distributor has the right to allow this—the distributor who stands to lose everything if it happens.

Here's another scenario: imagine for a moment that every book you bought at Walmart could be shelved only on a bookcase from Walmart. The books were designed with some kind of little divot in the spine, so that they could sit flat only on a Walmart shelf, and printed with a special ink that glowed only under a Walmart light bulb, and only when held at a special angle you could attain only by sitting in a specially designed Walmart chair. Every time you sold a ten-dollar book through Walmart, that would be ten dollars' worth of investment in this Walmart ecosystem your readers would feel beholden to, even if you and all their other favorite authors were later offered a better deal at Barnes and Noble (meanwhile, you'd only get fifty cents of that ten dollars in royalties, if that!). It's easy to understand why Walmart would love this—it creates a winner-takes-all market, where a small advantage quickly grows into an unbridgeable gap. The question is, why should authors or publishers want to have anything to do with a scheme like this?

Digital locks are roach motels: copyrighted works check in, but they don't check out. Creators and investors lose control of their business— they become commodity suppliers for a distribution channel that calls all the shots. Anti-circumvention isn't copyright protection: it's middleman protection.

Hachette, one of the largest publishers in the world, learned this the hard way in 2014. Amazon demanded a deeper wholesale discount from Hachette, and Hachette declined. Amazon retaliated by blanking out many Hachette titles on its site (including books by best sellers like J. K. Rowling), either marking them as either unavailable or on back order, and even suggesting books by other publishers, or used copies of the Hachette titles, as alternatives.

Hachette—the most fervent digital-lock advocate in the publishing world—was reduced to offering scolding comments in retaliation.

If it hadn't been for the locks on the Hachette titles, the company would have had a much more fearsome weapon at its disposal, when it came to its Kindle titles: it could have offered a 10 percent "Amazon Refugee" discount at Google, Barnes and Noble, and other retailers, and posted a free tool like the Calibre reader so that existing Amazon customers could easily convert their purchases to run on competitors' platforms.

But under the DMCA, only Amazon can authorize the conversion of Kindle books to read on non-Kindle platforms. Good luck with that, Hachette.

Platform as roach motel

Brewster Kahle is a bit of a software legend. He created the first search engine, the Wide Area Information Server (WAIS), sold it, founded another search company, Alexa, sold it, and then decided to spend the rest of his life running the Internet Archive (archive. org), an amazing public library for the Internet. Brewster tells a famous story about life in the shadow of Microsoft during the heyday of the packaged-software industry, when all software was sold in boxes hanging from pegs in software stores.

Back in those days, Microsoft owned 95 percent of the operating-system market, and spent a lot of time extolling the virtues of its "platform" (Windows) to its "partners"—the software creators who wrote Windows programs. The software vendors' associations also spent a lot of time warning programmers about the risks of piracy, and the problems they'd have if their customers decided to copy them into the poorhouse. This served to drive still more creators to the Windows platform, which was meant to provide a shield against this threat.

But then Microsoft started to undercut the same companies that had trusted its platform. In my hometown of Toronto, Delrina built itself into a huge, successful company by selling a ubiquitous fax program, WinFax, right up to the day that Windows 95 shipped with a free competitor, straight from Microsoft. Delrina, seeing which way the wind was blowing, sold itself to Symantec.

As Brewster tells it, Microsoft spent years waving one hand over its head, shouting, "Look out! Pirates! Pirates!" In the other hand, it held the knife.

Amazon MP3s: a lateral move

What happened to digital locks for music, anyway? How did that get to be the only entertainment category to completely abandon them? Funny story.

In the beginning, there were a bunch of "proprietary" music stores that signed up one or two labels each, and distributed music using their own digital locks. No one liked these stores very much. Then Apple created the first iPod, in 2001, and went around to the labels and said, "Look, these things only work with Macs, because they require an obscure cable-port called Firewire. That's only 5 percent of the market. Why don't you try an experiment with us?" And the iTunes Store was born, with music from all the big labels. The experiment turned out to be successful—so successful that the labels didn't dare pull out when the iPod started to work with Windows as well.

But Apple had lots of rules about selling music. You couldn't buy the kind of promotion that the labels wanted, and you could sell songs for only one price: ninety-nine cents. The labels hated this. Apple wouldn't budge on it. The labels came to realize that they'd been caught in yet another roach motel: their customers had bought millions of dollars' worth of Apple-locked music, and if the labels left the iTunes Store, the listeners would be hard-pressed to follow them. Just to make this very clear, Apple threatened a competitor, RealNetworks, when Real released a version of its player that allowed users to load (digitally locked) songs bought from the RealPlayer store onto an iPod, enabling customers to play both Real's and Apple's music on the same device. "We are stunned that RealNetworks has adopted the tactics and ethics of a hacker to break into the iPod, and we are investigating the implications of their actions under the DMCA and other laws," Apple said.

But Amazon offered the labels a lateral move: give up on digital rights management (DRM) software and sell your music as "unprotected" MP3s (which also play on iPods), and you can start to wean your customers off the iTunes Store—or at least weaken its whip-hand over your business. You can set your own pricing, Amazon said; we'll help you with the promos you're looking for, and together we can get some competition into the market. The music industry bought into it, and iTunes dropped DRM not long afterward.

Of course, book publishers totally missed this, and happily turned their catalogs over to Amazon to lock up with its proprietary Kindle format. Yikes!

So Is This *Copy* Protection?

WELL, IF ANTI-CIRCUMVENTION hands the keys to the business to digital-lock companies, at least it stops people from making infringing copies, right?

Nope, sorry.

There's an Internet meme called "Now you've got two problems." As in: "Your shower broke, and you decided to fix it yourself instead of calling a plumber. Now you've got two problems."

That's just what's about to happen with our digital-locks explanation.

You wanted to stop people from copying your file, so you gave them a ciphertext version and a player with a key hidden in it. But that player decrypts the file, and if your audience can save the file while it's decrypted, then they can start to copy it.

And now you have *three* problems, because they can also just take your keys. If they can find those, they can throw away your special player and make their own.

And now you have *four* problems, because users who figure out how to get the keys out of your player can tell other people how to do that, too.

And now you have *five* problems, because they can also just publish the keys, which are small and exceedingly difficult to suppress. Blu-ray's keys are 128 bits long—you could spraypaint one of them onto a smallish wall.

Is Apple for or against digital locks?

One of the world's most successful digital-lock vendors is Apple. Despite public pronouncements from its late cofounder, Steve Jobs, condemning DRM, Apple has deployed digital locks in nearly every corner of its business. The popular iOS devices—the iPod, iPhone, and iPad—all use DRM that ensures that only software bought through Apple's store can run on them. (Apple gets 30 percent of the purchase price of such software, and another 30 percent of any in-app purchases you make afterward.) Apple's iTunes Store, meanwhile, sells all its digital video and audiobooks with DRM. Many people assume that this is at publishers' insistence, but it's not so: when Random House Audio published the audiobook of my novel *Little Brother*, Apple refused to carry it without DRM.

Digital Locks Always Break

DIGITAL-LOCK VENDORS TEND to focus on how hard their technology is to beat if you attack it where it's strongest. "We have 256 bits in our key, and it would take a trillion years to guess it," they say. But this doesn't matter if you can decrypt the file without having to *guess* the key—if, say, you can find the key just by looking hard inside the player for it, or if you can find a way around using the key entirely.

There's a little joke about the guy who drops his keys by his car but searches for them under a lamppost a ways off: when someone asks what he's doing, he says, "This is where the light is." Digital-lock vendors armor the parts of the file they know how to armor, and then cross their fingers and hope that the people who want to copy their files don't attack on their weak flanks, even though there's no conceivable reason for this not to happen.

Take, for instance, e-books. Never in the history of the world have there been more qualified typists than there are today. My grandmother, a secretary, could hammer out seventy words per minute, and she was regarded as a kind of circus performer for it by the "girls" at the office. Today, seventy-word-per-minute typists are as common as air. Walk into a tenth-grade computer lab and close your eyes and you

How to kill the book

Copyright treats all creative works the same way. Needlepoint patterns and crossword-puzzle clues get the same protection that video games and blockbuster movies do. But books aren't like other media—books are *old*. Much older than copyright. Older than property. Older than markets. We have a particular reverence for books in our society, one that borders on superstition. If you were making a movie and you wanted to demonstrate that our world had been reduced to barbarism, you could just show a gang of angry townsfolk burning some books. Destroying a book has the same emotional tenor as eating a dog. After all, both books and dogs have been loyal companions and indispensable servants to the human race for millennia.

We know what the traditional book bargain is. Books can be shelved. Treasured. Lent. Passed on. Books belong to the people who acquire them, but they are also a responsibility, something to be curated and looked after. This sentimental attachment is of incalculable value to the publishing industry, which sells innumerable books to people who merely want to display them for status, or who feel that owning books is "the right thing to do."

But a digital lock on a book says that you're not the book's owner, merely its licensor, whose rights are set out in a long, incomprehensible "license agreement" you have to click before parting with your money. Most people understandably pretend that this doesn't matter, and that the ownership terms of books continue to be bound by the social contract that stretches back to the Roman Empire. Good thing, too: if publishers *do* succeed in convincing readers that books are just like CDs and DVDs, then the sight of a book in a shredder will have no more emotional kick than the sight of an old, scratched CD in the gutter, and publishers will lose all the social value they currently get for free.

could almost believe you were standing in a tin shack in a rainstorm—all those talented young fingers hammering at the keys, the telltale *whack* of the keyboards like the *pings* of raindrops bouncing off a roof. Turning the printed or digitized words of a book into a text file of a book is a tedious chore, but there are plenty of people out there who are up for it. And once someone has done it, the book can be shared instantly.

In 2007, J. K. Rowling, who had retained (and never exercised) e-book rights to each of her Harry Potter novels, released the seventh and final volume in the series exclusively as a print edition. Fans responded by converting it to an e-book within twenty-four hours, first lining up to get their copies at bookstores at midnight, when they went on sale, then retyping or scanning every page. Meanwhile, in Germany, another group of fans, impatient for the official German edition, took it upon themselves to *translate* the entire book into German. A few days later, their German edition was in hand.

Mind you, *Deathly Hallows* wasn't a little book. It stands (or, rather, falls crashingly to the floor) at a whopping 784 pages. And it's not as if such feats have gotten harder since 2007, nor are there fewer people qualified to perform them. As I once wrote in a novel, "Never underestimate the determination of a kid who is cash-poor and time-rich."

Now, maybe you don't fancy retyping a book. There are other ways. Google's Book Search program scanned books and converted them to text at such a furious clip that fifteen million books were digitized in six years—this being nearly every book ever published. Building your own home-brew book-scanner costs less than three hundred dollars, and there are, on the internet, extremely well-written weekend-project instructions for how to do it. How convenient.

Perhaps this still seems like too much work. Start with the e-book, then—all you need to do is download a free screen-capture program, one that is capable of capturing a predetermined region of your screen at the click of a button. Pair it up with your e-book-reading app (Amazon's Kindle app, say), click the button that takes you to the first page, and then click the button that captures and saves the rectangle of screen where the page is. Do this once for every page in the book—call it one page per second—and you'll end up with a folder full of pages.

Now upload those pages to Google's free optical character-recognition software (which converts pictures of words back into plain text), download the results, and call it a day.

There are analogs to these processes for practically all locked media. You can play locked audio out the headphone jack of one device and into the mic jack of another, recapturing the audio. You can plug the high-definition analog outputs from your media player into the high-definition analog inputs on your computer and recapture video.

All of these methods are admittedly tedious. If you really wanted to go into the lock-breaking business, you'd have to know something about programming.

The truth is that most digital locks that anyone cares about are broken in a day or two. In 2006, a programmer who called himself

Muslix64 broke the locks on a software-based HD-DVD player. He chose the player because it was the easiest one to break—that is, it was the one with the worst security. Remember that: people who attack digital locks get to pick their targets, and they generally choose the easiest one.

Back to Muslix64. He reasoned that, while he was playing an HD-DVD movie, the encryption keys must be somewhere in his computer's memory. So he put on a movie and looked through the memory until he found the key he wanted. After he announced his success, another programmer, arnezami, used Muslix64's methods to find a key that could be applied even more widely: the "09 F9" processing key, which can be written out as 09 F9 11 02 9D 74 E3 5B D8 41 56 C5 63 56 88 C0. That's hexadecimal, one of the counting systems used by computer programs. In decimal, the number looks like this: 13,256,278,887,9 89,457,651,018,865,901,401,704,640. Once this secret was released onto the net, any skilled programmer could write a program that descrambled HD-DVD movies; if she so chose, she could include the option to convert HD-DVDs to unscrambled video files that could be played without restrictions. No more piracy warnings. No more ads. And the DVDs you bought could be played anywhere in the world not just on players that came from the same country as the disc.

Muslix64's story is instructive—he attacked the weakest point in the system, and once he did, the whole system failed. There was a weird Keystone Kops moment when the big studios and their technology partners tried to prevent publication of the 09 F9 key—every time they did, it made the news, and affronted Internet users republished it somewhere else. Soon, there were millions of copies of the number. As of July 2014, Google knew of about 518,000 web pages containing the "secret" number.

Other keys have been discovered since then. Programmers routinely turn these extracted keys into programs that automatically strip the locks off of digital files, so you needn't be able to take a lock off yourself in order to accomplish this feat. When a programmer writes a lock-breaking program, she effectively bottles her expertise and sends it to you.

Still, many of these lock-breaking programs are written with a technical audience in mind, and are pretty tricky to get working. (Though some, like the DVD-ripping program HandBrake I mentioned earlier, rate with any commercial software for ease of use.) But chances are someone has already taken care of that, too—each new key is used to decrypt countless files by knowledgeable audience members eager to liberate their media before the enforcement catches up. You can just search the Internet for a copy of the movie, book, game, or song you want—it's a safe bet that someone, somewhere, has already unlocked it and put it out there for you to download.

Or, as they say on the Internet: now you have 518,000 problems.

Speed bumps and "honest users"

If you ask companies why they bother with digital locks when they're so easy to overcome, they say things like "Speed bumps don't prevent you from speeding, but they make it harder to speed. This is a digital speed bump." Unfortunately for them, the evidence doesn't bear out the comparison.

BigChampagne is a media-metrics company that produces usage statistics for file-sharing systems. Back when iTunes required digital locks on all its music, the BigChampagne folks decided to time how long it took for a song to go from iTunes to the file-sharing underground. They chose a song that had an early, iTunes-only release (to exclude songs that were being ripped from unlocked CDs), and measured away.

They reported that the net time between the release of a song with a digital lock and the appearance of the same song, unlocked, on file-sharing services, was *three minutes*. This 180-second gap represented the maximum time that a potential customer would have to wait before he could get a free copy of the song in question, complete with all the freedoms and permissions that had been locked out by iTunes.

The rejoinder to this from digital-lock advocates is that even if their technology can't deter lawbreakers, it can still "keep honest users honest." If you *want* to do the right thing, the lock tells you when you're about to fail to do so. But as the eminent computer scientist Ed Felten (a professor at Princeton, and formerly the chief technologist of the Federal Trade Commission) once quipped, "Nothing needs to be done to *keep* honest people honest, just as nothing needs to be done to keep tall people tall." If this is all a digital lock can claim to do, then it's not doing very much at all.

Sometimes, things get even muddier. Every three years, the U.S. Copyright Office entertains proposals for exemptions to its digital-lock rules. In 2010, the Copyright Office granted Americans permission to break the digital locks on their iPhones (but not their iPods or iPads), ruling that "jailbreaking" a phone after purchase constituted "fair use." However, this permission doesn't legitimize the creation or dissemination of tools to accomplish the task—it only confers on each user the right to discover a vulnerability in their iPhone, figure out how to bootstrap that vulnerability into a full-blown jailbreak, and then use it. It doesn't permit you to tell anyone else about the approach you took, and it certainly doesn't permit you to jailbreak someone else's iPhone or distribute a tool that does it for them. It's the equivalent of legalizing heroin use, but not production or sale. If you can magically make heroin appear out of thin air, you're allowed to inject it. Otherwise, you're still stuck illegally buying unknown substances from criminals and injecting them into your body.

Watermarking

Some digital-lock companies have promised to solve their problems with "watermarking." They propose to embed imperceptible signals in video and audio (and even text) that computers can search for, confirming that a file is in the hands of its rightful owner. A movie would thus be watermarked with its owner's information, saying, effectively, I AM CORY DOCTOROW'S COPY OF STAR WARS. If a DVD player that isn't registered to Cory Doctorow tries to play the movie, it's locked out. Or, alternately, companies could look on file-sharing sites for illicit copies of their movies, use the watermarks to figure out whom those copies originated with, and sue or blackball or otherwise punish that person.

Or maybe they could simply use the watermark to figure out which player was found to be weak enough to allow their locks to be removed, and future media could be made inaccessible to that model.

All these solutions are rife with problems. You can disallow some players, but you can't stop coders from writing and distributing their own playback programs. You'd also have a hard time finding a judge who'd convict someone of copyright infringement because the media originally sold to them was now available elsewhere. You'd have to prove that they released the files themselves, rather than, say, that they lost their laptop. Or sold it. Or lost it in a divorce. Or gave it to their children. Or had it stolen.

Finally, there's the big technical problem: a watermark *can't* be truly imperceptible unless you also want it to be useless. Imagine a photo with the words COPYRIGHT CORY DOCTOROW written in letters that couldn't be seen without a powerful microscope. A copy of the photo without the copyright notice would appear functionally identical to the copyrighted version, unless you went to great lengths to inspect it. Of course, you could write COPYRIGHT CORY DOCTOROW across the photo in four-inch black letters, but you'd have a hard time selling a photo like that.

Watermarkers have tried to find a middle ground, but that leaves their work exposed. People who want to remove watermarks can just get two or more copies of a file, compare them byte for byte, look at the bytes that are different in each file, and scramble those bits. Job done. Ed Felten (see previous sidebar) once assembled a team of researchers that looked at the difficulty of removing the Secure Digital Music Initiative watermark—the product of the most ambitious and expensive watermarking project ever undertaken. They figured out how to do it with ease.

Understanding General-Purpose Computers

BACK AT THE dawn of mechanical computation, computers were "special-purpose." One computer would solve one kind of mathematical problem, and if you had a different problem, you'd build a different computer. But during World War II, thanks to the government-funded advancements made by such scientific luminaries as Alan Turing and John von Neumann, a new kind of computer came into existence: the "general-purpose" digital computer.

These machines arose from a theory of general-purpose computation that showed that, with a simple set of "logic gates" and enough memory and time, you could "compute" any program that could be represented symbolically. That is, general-purpose computers are machines capable of running every valid program we can write. This principle gave birth to the modern computer as we understand it. It suggests that all computers are equivalent on a fundamental level—any computational task can run on any computer. (Though some computers might take trillions of years to run the whole program.)

It's nearly miraculous, this business. Until the middle of the last century, most computers were purpose-built to compute in one particular way. Today, we can—and do—routinely make computers that can compute in *every* describable way. Our world is made up of general-purpose computers.

This is the major reason that digital locks have been dead on arrival since their first attempt, and why they continue to fail. We know how to build a computer that can solve one kind of problem (like a mechanical adding machine), and we know how to build a computer that can solve *all* kinds of problems. But we don't know how to design and build a computer that can run every program except for one program that pisses off, endangers, or harms the entertainment industry.

The computer that runs everything minus one has no basis in theory. Maybe some twenty-first-century Alan Turing will invent it, but no one has yet stepped forward with such a design, and the consensus in computer science is that such a design is not feasible. Which means that we can't deliver on the promise of digital locks.

Of course, that hasn't stopped the entertainment industry from trying to approximate the effect of a computer that doesn't run a certain program. And that's where the trouble begins.

Instead of building computers that are incapable of running a program that makes copies of files that shouldn't be copied, we make computers that are designed to run all programs. Then we add software that watches what that computer's user does at all times, and that secretly kills any process—any program—that does something the entertainment industry doesn't like.

As a rule, computer owners don't want their computers to disobey their orders. From the owner's perspective, a computer that won't make copies when you want it to is broken. So digital locks need to be designed for stealth as well as vigilance. They need to disguise themselves from users, and hide their workings from the operating system.

There's only one class of programs that behaves this way, hiding itself from users and doing things that users don't want: spyware.

Rootkits Everywhere

THE MAJOR STRATEGY behind both spyware and anti-copying software is called a "rootkit." Rootkits are programs that covertly modify a computer's operating system to blind it to certain files and processes. Once a computer has been compromised by a rootkit, it will not "see" the files associated with the malicious software the rootkit was designed to conceal. If you open a folder containing said malicious program on a rootkit-infected computer, the program won't be visible in the folder's listing. Nor will you be able to see any of the processes associated with that program: if you go to your computer's "process monitor" (the utility that lists all the programs running on your computer), the malicious program will not appear on the list, even if it's running.

It's useful to think of this in terms of your own immune system. A healthy immune system is capable of recognizing and categorizing the stuff in your body. When your immune system is working well, it can distinguish between helpful gut bacteria, toxic, infectious bacteria inhabiting a cut on your hand, and your own cells. Auto-immune disorders occur when your immune system gets confused. Sometimes your immune system will erroneously identify your own body as a harmful foreign agent, and attack it—causing anything from mild allergies to horrific and fatal outbreaks. On the other hand, sometimes your immune system stops being able to combat harmful organisms, and lets them multiply without interference—this is what happens when pneumonia brings down people whose immune systems have been compromised by HIV.

It's not a coincidence that spyware and digital locks use the same rootkit techniques. In 2005, Sony BMG shipped six million audio CDs loaded with a secret rootkit that covertly installed itself when you inserted one of those CDs into your computer. Once your computer had been compromised, any file or process that began with "sys"

was invisible to the operating system. So a file called "sysrootkit.exe" wouldn't show up when you looked in the folder containing it, and when that program ran, sysrootkit.exe wouldn't be listed in the process monitor.

The Sony rootkit was used to cloak a program that watched for, and then killed, attempts to copy music off of audio CDs. From the user's perspective, it looked like your computer had suddenly developed a mysterious bug that stopped CD-ripping software from running. This is the equivalent of an auto-immune allergy—the computer's owner has a process that she *wants* to run (a CD-copying process), but the computer erroneously identifies that program as undesirable, and kills it every time it tries to start.

But it didn't stop there. Once there were millions of computers in the wild that couldn't see files that started with "sys," virus writers started to add "sys" to the names of their programs, too. This was the digital equivalent of HIV: your computer could no longer detect bad programs, because a gaping hole had been punched through the middle of its immune system. Any program that adopted the protective coloration of Sony's anti-copying tool could cloak itself in Sony's rootkit.

Thus we arrive at yet another problem with digital locks. Fundamentally, the only way to get computers to resist users' attempts to see what's going on inside them is to design programs with the facility to misreport their files and processes. And forcing such programs on users inevitably opens the door to spyware.

What your PC sees

In 2009, a scandal erupted in Lower Merion, Pennsylvania, an affluent suburb of Philadelphia. The Lower Merion School District had used state grants to supply 2,300 high-school students with district-owned laptops. These computers came loaded with secret software that could operate the laptops' cameras covertly (when the software ran, the cameras' green activity lights stayed dark), as well as capture screengrabs and copy the files on the computers' hard drives. This software was ostensibly in place to allow the district to track down stolen laptops.

But that's not how it was used. Instead, a student discovered the software's existence when he was called into his assistant principal's office and accused of taking narcotics. He denied the charge, and was presented with a photo taken the night before in his bedroom, showing him consuming what appeared to be pills. The student explained that these were Mike and Ike candies, and then proceeded to ask how the hell the assistant principal had taken a picture of him in his bedroom at night. It emerged that this student, and a number of others, had been photographed *thousands* of times by their laptops' cameras without their knowledge: awake and asleep, clothed and undressed.

Appliances

A RECENT TREND in the computer industry is to sell general-purpose computers as "appliances" with limited functionality. In the last few years, companies have embedded computers into digital photo frames, home WiFi routers, portable music players, pacemakers, and car dashboards. This is inevitable: after all, general-purpose computers are made in enormous quantities, and boast ever-increasing speed and ever-decreasing price. Why make a special-purpose chip that can only tell the time or play an MP3, when there's an off-the-shelf alternative that can do all that and more, available at a fraction of the cost?

I love my digital appliances—my mobile phone, my bedside clock, the router that runs my home broadband, the waterproof music player I use when I swim. But the availability of general-purpose technology has tempted some companies into digital-lock territory, and that's where it starts to go sour.

Take home broadband routers. Many companies would like to offer different models of the same router, with varying features—the basic model that does just WiFi, a more deluxe version that can also serve as a voice-over-IP telephone, and a top-of-the-range version that can do the kind of sophisticated routing tricks used by major corporations and universities to manage enormous, complex networks. The thing is, it's often cheaper to base all three models on exactly the same hardware, and simply load different software at the point of manufacture to make the products different. This is innocuous, in and of itself. However, it leads to something awfully ugly.

Once a company stands to lose money by allowing customers to examine and modify its products after they're sold—by switching out that software, say—it creates a perverse incentive to treat customers as threats, and to try to hide the workings of their devices from them.

This means that digital "appliances" are very different from their analog cousins. An electric motor is useful in both a hand drill and a blender. Once you install the motor in a blender, it becomes difficult to use it in a hand drill, but there's nothing stopping you from trying. You can use your blender in ways the manufacturer never thought of—you can mix paint with it, or take it to your high-school lab for experiments, and while it's true that it would take an extraordinary measure to turn your blender into a hand drill, that's not because the manufacturer has spent good money to stop you from doing it. And if you figured it out, it wouldn't be against the law to do it, or to tell others how you did it, or even to sell blender-drill conversion kits.

Contrast that with digital appliances. A digital appliance isn't a computer that's had all its features removed so that it can run only the "appliance" program; instead, it's usually a fully functional general-purpose computer capable of running *every* program, but shipped with spyware and other countermeasures installed so that it comes out of the box treating you, its owner, as an attacker, and does everything it can think of to stop you from seeing its inner workings. What's more, if you *do* figure out how to get inside the machine and make it do more than the manufacturer intended (for example, if you jailbreak an iPad), you're very likely breaking the law, and you're also prohibited from telling anyone how you did it or from going into business doing it yourself.

Proto-Appliances: The Inkjet Wars

AN EARLY EXAMPLE of the "adversarial appliance" model is the inkjet printer. Official, licensed inkjet ink is just about the most expensive liquid you can buy, costing more per ounce than vintage Champagne. That's because the business model of many printer companies has been to sell printers cheaply—sometimes even at a loss—and then to make their profits on the ink refills.

This works only if users can't refill their cartridges themselves, or use a competitor's ink. If HP is marking up its colored water at 100,000 percent margins, you'd expect a competitor to come along and offer the same product at a more modest 1,000 percent margin, or even less. But HP protects its margins by designing printers whose inner workings are kept secret from the owners of the hundred-million-odd HP printers in the field today. No one is supposed to know how HP's operating systems work, because that would help them figure out how to trick HP printers into accepting refilled, remanufactured, and third-party ink cartridges. Every time that happens, HP loses out on a hugely profitable cartridge sale.

This secrecy isn't harmful just to HP printer owners' wallets—it's harmful to their security. In 2011, Columbia computer-science grad student Ang Cui conducted research into the security of HP printers. HP refused to disclose the inner workings of its printers to him, citing commercial confidentiality, so Cui undertook to reverse-engineer their technology himself, and was able to quickly unravel the system. He found that HP had devoted a lot of resources to preventing the use of refilled cartridges, but almost none to other types of security.

To demonstrate this, Cui wrote a simple two-hundred-line program that could turn any document into a vector for hijacking HP printers. If Cui could convince you to print his document (if, for example, he sent a résumé to your company's HR department, and they printed it), he could seize control of your printer. Once the printer was under his control, it would no longer accept software updates from its owner, though it would *pretend* to accept them, and register them as being successfully installed. It would also spy on every document printed, and send copies to Cui's personal drop box. It could even look for certain words or numbers in documents (say, Social Security numbers) and alter them when the documents were printed.

Cui's program also scanned the local network for PCs that had known vulnerabilities that hadn't been patched with a software update. Once it found these computers, it took *them* over, installed rootkits on them, and used them to penetrate their network's firewall and open a connection to Cui's laptop. In other words, it broke the whole network wide open.

Cui is a very clever security researcher, but he's not a transcendent genius. It's a near certainty that other people figured out these techniques independent of Cui, and the fact that we never heard about it likely means that they kept them to themselves—or sold them to the highest bidder in the international identity-theft and espionage markets.

Without a thorough understanding of our computers' workings, and without independent verification of their security, it's impossible to trust our machines. Adversarial appliances require that computers' workings be kept secret, and digital-lock rules impose civil and criminal penalties on people who discover and publicize their vulnerabilities. That's a big problem.

Carrier IQ, Android, and iOS

In 2011, a security researcher named Trevor Eckhart discovered that some Android phones had shipped with a commercial rootkit called "Carrier IQ." Carrier IQ bills itself as a tool for monitoring and improving mobile-network performance by gathering statistics about dropped calls, signal strength, and so on. But as Eckhart demonstrated, Carrier IQ could do much more: it could record your passwords as you entered them on your phone's keyboard, read your text messages, and pinpoint and transmit your location. When Eckhart published his findings, the carriers performed the classic deny-threaten-downplay tactics, first denying that they'd done anything wrong, then threatening to sue Eckhart, then, finally, sheepishly admitting their wrongdoing and apologizing. As this little soap opera was unfolding, the iPhone hacker known as Chpwn used what had been discovered about Carrier IQ to go hunting for it on iPhones, which run iOS, the closed-up rival to the (mostly) open Android operating system.

Chpwn was able to ascertain that Carrier IQ was present on iOS, too. Ultimately, it emerged that over 141,000,000 handsets in the USA had been infected with the Carrier IQ rootkit by the carriers themselves, without the knowledge or consent of the phones' owners.

The important lesson here—apart from the fact that telcos are lying, evil scum—is that it's easier to spot shenanigans in open systems than in closed systems. It's not a coincidence that the Android revelation came first, because just starting up an iPhone with an OS other than the one provided by the carrier and manufacturer is illegal, and security researchers who publish material about the iPhone's workings face legal jeopardy, as do their publishers, employers, and academic institutions. The digital-lock rules in the DMCA make it illegal to figure out how to install your own OS on an iPhone, and prohibit publishing any information that would help someone else do the same. Developers—who are allowed some limited transparency on the inner workings of iOS devices—are required to sign nondisclosure agreements as a condition of access.

Imagine a home-security company whose burglar-detection system let them secretly listen in on your private moments at home and watch you through your CCTV cameras. Now imagine that it was illegal to tell anyone about the way that system worked. That's the world we're living in.

Worse Than Nothing

Back to the WCT

At the start of this chapter, I wrote that the WCT created the digital-lock rules that the nations of the world turned into domestic law, with small variations from state to state. But the truth is, most nations have followed the lead of the U.S. in going much further than the WCT requires. The WCT calls for laws against breaking a lock to commit an act of copyright infringement, but the DMCA, here in the U.S., makes it illegal to break digital locks, period. Other countries around the world have followed suit, often at the open, explicit urging of the Office of the U.S. Trade Representative and lobbyists from the U.S. entertainment industry. Which means that the problem is even worse than the WCT suggests.

THE TECHNICAL IMPLAUSIBILITY and unintended consequences of digital locks are big problems for digital-lock makers. But we're more interested in what digital locks do to creators and their investors, and there's one important harm we need to discuss before we move on. Digital locks turn paying customers into pirates.

One thing we know about audiences is that they aren't very interested in hearing excuses about why they can't buy the media they want, when they want it, in the format they want to buy it in. Study after study shows that overseas downloading of U.S. TV shows drops off sharply when those shows are put on the air internationally. That is, people just want to watch the TV their pals are talking about on the Internet—they'll pay for it if it's for sale, but if it's not, they'll get it for free. Locking users out doesn't reduce *downloads*, it reduces *sales*.

The first person to publish a program to break the digital locks on old-style DVDs, in 1999, was Jon Lech Johansen, a fifteen-year-old Norwegian teenager. "DVD Jon" took up the project because his computer ran the GNU/Linux operating system, for which the movie studios wouldn't license a DVD player. In order to watch the DVDs he bought, he had to break their locks. Seven years later, Muslix64 broke HD-DVD's DRM for similar reasons—he wanted to watch a legitimate out-of-region DVD that he'd purchased. Both of these seminal figures in the history of digital locks were inspired not by "piracy" but by frustration with the limitations put on the legitimate media they'd paid good money for.

In 2007, NBC and Apple had a contractual dispute over the terms of sale for Apple's iTunes Store. NBC's material was withdrawn from iTunes for about nine months. In 2008, researchers from Carnegie Mellon University released a paper investigating the file-sharing impact of this blackout ("Converting Pirates Without Cannibalizing Purchasers: The Impact of Digital Distribution on Physical Sales and Internet Piracy"). What they found was that the contract dispute resulted in a spike of downloads on "pirate" sites, and not just of NBC material—it seemed that once people who had been in the habit of buying their shows on iTunes found their way onto the free-for-all file-sharing sites, they clicked on everything that looked interesting. Downloads of NBC shows went up a lot, and downloads of everything else went up a little.

More interesting is what happened after the NBC-Apple dispute ended, and the shows returned to iTunes. As the CMU paper showed, download rates for those shows *stayed* higher than they had been before the blackout. That is:

- Refusing to sell viewers the content they wanted in the format they preferred drove those viewers to piracy.
- Once the audience started pirating the content they wanted, they quickly turned to pirating other content, too.
- Having become aware of and proficient in the ways of downloading, the audience developed a downloading habit that outlasted the end of the blackout.

Digital-lock vendors will tell you that their wares aren't perfect, but they're "better than nothing." But the evidence is that digital locks are *much* worse than nothing. Industries that make widespread use of digital locks see market power shifting from creators and investors to intermediaries. They don't reduce piracy. And customers who run into frustrations with digital locks are given an incentive to learn how to rip off the whole supply chain.

If you're a publisher, label, or studio, the answer is simple: don't let companies sell your goods with digital locks on them. And if a company refuses to sell your goods unless they can put their locks on your products? Well, you can be pretty sure that those locks aren't there for your benefit.

It's harder if you're a creator, because many of the biggest investors have bought into the idea of selling with DRM or not at all. When it comes down to negotiating DRM, you just have to make a decision about whether you're willing to let your creative work be put in some tech company's jail in order to make your investors happy, or whether you'll keep shopping for a saner, better investor.

A few years back, I sold a children's picture book to the largest publisher in the world (which will remain nameless here), which then spent a couple of years developing it with me, commissioning rough illustrations and going through several rounds of rewrites until it was something we were all excited about.

One of my editors at Nameless Giant Publishing was also the head of its UK digital strategy, and he and I were very sympatico, as you might expect. After several months had gone by without a contract showing up at my agent's office, he called me up and explained what had happened.

He'd gone to the contracts people and requested a no-DRM guarantee for the digital editions of my books. None of us thought it would be a problem, since Nameless Giant is my publisher in several other formats, languages, and territories, and in every case, it sells my work without DRM.

But Nameless Giant had a new directive, from the very top of the business. From now on, all books had to be acquired with e-book rights, and all e-books would have DRM on them. My editor tried to negotiate ("Can we not acquire the e-book rights? No? Okay, how about we acquire them, but promise not to use them?"), but it was in vain.

Finally, my editor explained to the contracts person that the expected return from the digital edition was -£80—that's *negative eighty pounds*. In other words, the company expected to *lose* eighty pounds on the digital edition, based on the performance of its other digital picture books.

The contracts person told my editor that the DRM was nonnegotiable, and that if it was going to be a problem, he should cancel my contract.

So my editor quit.

It's very hard to get angry with an editor who has just quit in protest over your book getting canceled, even in the face of your book getting canceled.

The story has a happy ending. Nameless Giant sank thousands into developing a book that I can now easily sell to one of its less doctrine-blind competitors, and I don't have to let DRM companies lock up my copyrights. Also, my editor got a much better job.

Fame Won't Make You Rich, But You Can't Get Paid Without It

SO DIGITAL LOCKS can't stop copying—in fact, they only make things worse. So how *should* artists think about unauthorized copies of their work? Can we just treat copies as advertising?

Well, yes and no.

Tim O'Reilly is the founder and publisher of O'Reilly Media, one of the largest tech-book publishers in the world. O'Reilly Media puts on some of the world's best-loved technology conferences, and generally serves as a hub of technology know-how for people all over. Tim O'Reilly is known for many things: he cowrote one of the first UNIX manuals, he popularized the term "Web 2.0." He's also famous for saying *"Obscurity is a far greater threat to authors and creative artists than piracy."*

I happen to agree with Tim on this point, but I worry that many folks who read those words decide that what Tim is saying is "Once you're famous, you'll be rich." If you read closely, you'll see that this isn't what he's saying. He's saying something like "Artists need to worry about fame before they worry about fortune." Recognition is one of many necessary preconditions for artistic success: luck, talent, and an indefatigable drive to succeed that lasts through the years and years it takes to get noticed, build a following, or get onto the radar of an important promoter, gatekeeper, or investor are a few of the

Does copying benefit anyone besides famous creators?

The copyright debate is filled with lies, damned lies, and piracy statistics. Figuring out whether unfettered Internet copying has helped or hurt an artist is a challenge—if you're trying to compare actual sales to sales in an imaginary world where it's possible to control copies, you have to agree to a lot of other assumptions. How many other imaginary-world artists are competing for your artist's audience's attention? Does the Internet still somehow exist as a promotional medium (without allowing uncontrolled copying)? Are we going to imagine that Top 40 radio is still the powerhouse it once was? Are we going to assume that people are still buying digital products—MP3s from Amazon, videos from iTunes, games from Steam—or are we going to assume that these things disappear, too?

But one charge is that being copied helps you make money only if you're already famous. Fans sent voluntary payments to

Radiohead for their pay-what-you-like smash-hit album *In Rainbows* only because a label had already spent a fortune building Radiohead's following for them.

Let's be clear: no one's going to give you money if they don't know you exist, so being famous (however you got there) will increase the likelihood that you'll get some money for your work. But artists have also found avenues to fame *and* fortune by giving stuff away—Jonathan Coulton, the now-legendary nerd troubadour who is probably best known for his song "Still Alive," the closing theme to the Valve game *Portal*, is one such example.

While working as a programmer, Coulton undertook to write, perform, record, and post a song to his website as a free podcast every week for a year. He also sold the songs as DRM-free MP3s, and encouraged his fans who didn't need the MP3s to consider donating to him through PayPal. ("Already Stole It? No Problem," his website says, before pointing to a PayPal link.) Coulton called the project "Thing a Week," and he acquired quite a following through it. Midway through the year he went out on tour, selling CDs and related swag. As his following grew, so did a number of fun and lucrative commissions—the *Portal* song that catapulted him to still-greater recognition being one of them.

others. Further, you can have all those things without being a "commercial success" through the simple expedient of being a commercial dunderhead who spends money faster than he makes it.

So, yeah, being famous won't—in itself—make you rich. But if nobody knows about your work, nobody's going to buy it.

The thing is that nearly everything that nearly everyone tries to get rich from a career in the arts fails. The major record labels, TV and movie studios, and publishers freely admit this. It's the reason they take such a big slice of the price of our media, relative to the creator's share—they have to invest in a *lot* of failures to get one "success." Looking at it that way, we can enumerate a few people for whom free copying has worked, and a lot of people for whom it hasn't worked. And we can name a few people for whom controlling copies—in the pre-Internet era—worked, and lots for whom it failed.

Fame isn't money. You can't pay for a plane ticket with fame. You can't pay for your kids' braces with fame. You can't pay for a copy of this book with fame. (Unless you're famous as a reviewer, in which case you can.)

But if you're in the arts, you'll never get money without some kind of fame. People can't give you money for your art unless they know it and you exist. They might *still* decide not to give you money for your art at that point, but without fame, they don't even have to make the decision. The Internet, fundamentally, is a thing that gives them a chance to do that.

Does this benefit anyone besides obscure creators?

If obscurity is a bigger problem than piracy, does that mean that obscure artists are the only ones who stand to benefit from the fact that digital locks can't last on the Internet?

Not at all. It just means that for famous people, *obscurity* isn't the problem that the Internet helps to solve. The Internet does more than make obscure things famous, after all: it also solves many other parts of the money puzzle. For example, the Internet makes it possible for a famous creator to directly distribute copies of his work to his audience without spending everything he brings in on marketing. That's what Louis C.K. did in late 2011—when he was arguably the hottest comic in the world. He paid about thirty-two thousand dollars to develop a website and about a hundred and seventy thousand dollars to make a professional recording of a show with all-new material, then put that video online for five dollars, without DRM. He made over a million dollars in the first two weeks. After his expenses and the transaction fees charged by PayPal, he had about seven hundred fifty thousand dollars of that first million left over. He proceeded to split it between himself, his staff, and five separate charities. C.K. showed that the absence of digital locks wasn't a disincentive to pay him. By performing an act of public generosity and trust, he inspired a million bucks' worth of reciprocal generosity and trust.

Good at Spreading Copies,
Good at Spreading Fame

THERE'S NO GOOD way to stop the spread of information on the Internet. If you've ever been unfortunate enough to attract a kook-enemy who posts all kinds of crazy conspiracy theories about you on the web, you'll have discovered that once the stuff is out there, it's very difficult to remove. Indeed, making a big stink about objectionable material often attracts *more* attention to it, because everyone wants to know what's so salacious that they shouldn't be allowed to see it. Mike Masnick of Techdirt coined the term "Streisand effect" to describe this phenomenon, after Barbra Streisand objected to an aerial shot of her house being included in a survey of coastal erosion in California. She sued the photographer for fifty million dollars, and in the process turned a picture that virtually no one had ever seen (the photo had been downloaded six times, before the lawsuit) into an icon of Internet photography.

Today, if you find yourself in Ms. Streisand's shoes, you might be tempted to turn to Internet "reputation managers" who'll charge you money to try to make the unwanted information disappear. They tend to employ a combination of search-engine tricks, heavy-handed legal threats, and intimidation. Save your money; most of those guys are ripoff artists. If the record industry and the movie industry and the TV industry and the publishing industry can't stop information from being disseminated, what are these pipsqueaks going to do?

We can't stop copying on the Internet, because the Internet is a copying machine. Literally. There is no way to communicate on the Internet without sending copies. You might think you're "loading" a web page, but what's really happening is that a copy is being placed on your computer, which then displays it in your browser. (Actually, in the case of a web page, there are probably *hundreds* of copies made in response to your click. You click and the page you requested is copied to server RAM, to routers, to caches, to more routers, to your network router, to your computer's network buffer, to your computer's drive cache, to your computer's RAM, to your computer's video buffer, and then to the screen.) The entertainment industry sometimes tries to argue that "streams" aren't copies, as in "You can buy the stream for a dollar, or download a copy for five dollars." But there is no such thing as a "stream" in this sense of the word. Streams *are* downloads. Your computer downloads the stream from a media server, and makes several more copies of the data before you get to see or hear it. Then the streaming software deletes the data before it can be saved to your hard drive.

This is a lot less neat than "You can buy a copy for five dollars, or you can just look at the movie once for one dollar." Every pixel of every frame that you watch, every sound wave in that song you listened to, spent some time on your computer. The difference between "streaming" and "downloading" is whether your program gives you a "Save As" button.

But all is not lost. As Barbra Streisand, Jonathan Coulton, Louis C.K., and innumerable others have discovered, the Internet isn't just a copying machine, it's an *audience machine*.

An Audience Machine

TECHNOLOGICAL INTERMEDIARIES LIKE online stores and distributors have used digital locks to shift market power from creators to themselves. These intermediaries were able to lock in, and hence own, the audience for the work they have for sale.

But intermediaries aren't one-dimensional villains. They act as gatekeepers, deciding whom to exclude and whom to admit, but they do so because they think they know what the audience wants.

It's not surprising that creators have a love-hate relationship with intermediaries. If Blockbuster Video—once the largest chain of video-rental stores in the world—decided that your movie wasn't up to snuff, or was too risqué for Middle America, your movie would, for all intents and purposes, die. On the other hand, if Blockbuster *loved* your movie, and featured it prominently in its marketing and in-store displays, if its clerks talked it up to everyone who came through the door, you'd have it made.

In some ways, digital-lock laws have reproduced that situation on the Internet. Once a market is established by a few digitally locked-in players, those intermediaries become increasingly powerful, and can flex their muscle to creators' and investors' detriment or benefit.

But the Internet has also *weakened* the power of intermediaries by increasing the number of ways that audiences can be united with creative works. When there's only one cable operator in town, that cable operator can call all the shots; add a satellite provider, and creators may get a slightly better deal (though it's amazing how quickly two competitors can independently arrive at near-identical abusive and nonnegotiable terms).

It's when you open the Internet to all the ways of connecting audiences to creators that things really start to change. Creators have never enjoyed a wider, more diverse, less united, and more pliable set of intermediaries than we have today. From YouTube to Twitter, Facebook to WordPress, Wikia to Tumblr and many, many (many, many, many, *many*) others, there have never been more ways for works and audiences to come together.

This is bad news if you're a success from the pre-Internet era, with a business model married tightly to the intermediaries who serve your markets. You might know to the penny what it will cost you to put a movie into theatrical distribution, or get a book into the endcaps in every chain store in the country. You're accustomed to being able to run a cost-benefit analysis: "A certain number of people will go to the movies every weekend. If I get one screen in every multiplex, I'll sell at least x tickets, and make y dollars."

In the age of disorganized and diverse intermediaries, you can still price out the endcaps, the payola for radio airplay, the theatrical distribution buyout. But you can't know what it's worth. There are lots of ways to buy books today that don't involve endcaps (or stores!). There are lots of ways to see movies today that don't involve cinemas or renting DVDs. You might still be able to guarantee that you'll get the lion's share of one channel or another, but you can't predict how much of the market will use that channel on any given day. Which makes it damned hard to grow your multibillion-dollar media empire by 3 percent quarter-on-quarter and keep your stockholders happy.

But if you're a creator who never got the time of day from one of the great imperial powers, this is your time. Where once you had no means of reaching an audience without the assistance of the industry-dominating megacompanies, now you have *hundreds* of ways to do it without them.

Of course, it bears repeating that reaching an audience isn't the same as convincing them that you have anything they want to see.

And if they do want to see (or hear, or read, or play) what you've got, they may not be willing to part with money to do it. But:

- No one can decide whether your stuff is worth money until they see it; and
- they can't see it until they know it exists.

Reaching audiences is a task of surpassing, almost mystical significance and difficulty. It's the problem shared by street-corner preachers, Madison Avenue mad men, spammers, and artists. Anything that makes it cheaper to reach a new audience necessarily increases the overall audience you can reach. And every new audience increase your chances of finding one that'll pay you.

Leverage

Generally, recording artists start off with a pretty rotten deal. The standard record contract gives control over "masters" (the master recordings, key to royalties and reissues down the line) to the label. Labels also deduct charges for "breakage" (physical merchandise that is damaged in shipping) from artists' sales, including from digital sales where there is no physical merchandise to be damaged. And labels bargain for a 7 percent royalty payout on "sales" and a 50 percent payout on "licenses," but class iTunes downloads and other digital transactions (which are licenses—according to iTunes, you don't own the iTunes music you buy) as sales, keeping 93 percent of the revenue from each ninety-nine-cent track, rather than 50 percent, which is what the actual contract says they should get.

And in an example of what can only be called theft, the labels used to run "third-shift" pressings of CDs in the dead of night, which were off the artists' books, and sell those CDs without any payment to artists. As noted copyright scholar William Patry documents in his book How to Fix Copyright, this routine industry-wide practice was ended only when the Sarbanes-Oxley Act made company executives criminally responsible for false accounting statements.

This rotten deal is largely non-negotiable, especially for new artists. Even famous and successful acts, who represent a major source of revenue for the labels, usually can't renegotiate the deals they signed starting out. A musician who has satisfied the terms of his initial deal and become a success can threaten to leave his label unless he gets a better deal the next time around, but he's not going to get control over his masters from the original contract. He's not going to get paid what he's owed on the iTunes sales of his last record.

The reason the deal is nonnegotiable is that it is industry-wide. There are only three major record labels, and they all offer the same rotten terms to their new artists. When you're the only game in town, you get to make up the rules, and tilt them to your benefit. It's a little like the old Lily Tomlin bit from *Saturday Night Live*: "So, the next time you complain about your phone service, why don't you try using two Dixie cups with a string? We don't care. We don't have to. We're the Phone Company."

Even for very successful artists, a new contract negotiation was always bounded by the "two Dixie cups and a string" ultimatum. As in, "None of the Big Three are going to give you what you're asking for. And if you don't like it, try getting your music to your audience without us."

Now, though, the explosion in Internet intermediaries has created a wild, fecund, and often confusing ecosystem of audience-reaching systems. There have never been more ways for audiences to be exposed to sound recordings, nor more ways for money to be made off this exposure. And when there's more than one game in town, the rules are up for negotiation.

That's where Radiohead's *In Rainbows* and Nine Inch Nails' *Ghosts I–IV* come in. These massive indie smashes changed the upper boundary of a successful artist's contract

renegotiation, because they showed that it's possible for a successful artist to make more money outside of the label system than in it. As a result, the labels were forced to open up their terms for their most successful artists.

But does that matter to someone who *isn't* a success? Sure, though not in the same way. Perhaps you've been gutting out a musical career for years, and, you're finally on the verge of getting a record deal. The existence of successful indie business models—and not just for creators who were made famous through the traditional system, either—means that the labels you negotiate with understand that you could plausibly go it alone.

It also means that there are a multiplicity of companies who *aren't* traditional labels who are in a position to offer you a better deal than you could get from the old take-it-or-leave-it label system. Ironically, some of the success of these companies is due to the labels' own business models. Decades of mergers and acquisitions across every part of the entertainment industry (print publishing, games, music, TV, and film) have led to massive layoffs of talented production staff, and a massive reliance on outside contractors who are mostly composed of ex-in-house talent—and who can now go to work for the indies, too.

In other words, the entertainment companies gobbled each other up, fired most of their staff, and now hire them back on a job-by-job basis as contractors. This means that in many cases, literally the exact same people are available for hire to spunky startups with disruptive business models. If you sign with an indie label today, there's every chance that you'll be recorded, mixed, and mastered by the same engineers and producers, in the same studio, that you would have been working with if you'd gone with a major. What's more, your album can be packaged, marketed, and publicized by the same people who'd have done that part of the job, too. The proprietary advantage once enjoyed by companies who assembled teams under their own roofs and used them only on their own products has been surrendered in the rush to attain some semblance of "lean, efficient" flexibility.

Streaming services

The latest poster children for the impact of leverage on artists' fortunes are the streaming music services—Last.fm, Spotify, Pandora, and the like. When Internet streaming started, the rules for paying royalties were unclear. Anyone could set up a streaming service so long as they escrowed some money for royalties, which would be paid once the rates were worked out.

In 2007, the Copyright Royalty Board—widely viewed as a creature of the Recording Industry Association of America and its member labels—handed down the new royalty system, and nearly all the streaming services disappeared. Hundreds of independently curated channels offered through services like iTunes, some hosted by famous DJs and musicians, shut themselves down. The new streaming rules made it nearly impossible for such services to exist—though they did allow radio broadcasters to set up streaming services at huge discounts.

Since then, a small handful of dominant streaming companies has emerged. These companies hand over gigantic sums of money to the music industry—far more per song than radio stations have ever paid—but nearly all of it is retained by the record labels, leading many musicians to accuse the services of ripping them off.

The reality is that streaming services offer copyright holders a much better deal than radio ever did. For one thing, radio pays royalties only to composers, while streaming services pay both performers and composers. That's right: musicians get *nothing* from U.S. radio play, and never have. Only the songwriters are entitled to a royalty. One major source of confusion is the relationship between a radio broadcast and a stream. A radio broadcast reaches an entire city or region, and has millions of potentail listeners; radio stations, nevertheless, pay a single rate for their whole audience. By contrast, *every person* who listens to a stream is accounted for separately. The per-play royalty from streaming services is thus much lower than the equivalent rate for radio—but that's because streaming pays per song *and* per person.

The problem with streaming isn't that it doesn't pay copyright holders.

It's that the labels have stacked the deck so that they are entitled to the lion's share of the money. They've done this by reducing the pool of streaming services to a tractable handful that could then be strong-armed into giving preferential treatment to the majors through direct deals that are opaque to musicians and subject to abusive accounting practices. In many cases, the majors now take huge income guarantees, or even equity, from the streaming services—money that covers their artist-royalty obligations and then some, leaving them with massive profits they're under no obligation to share with their musicians, and often don't. Merlin, a nonprofit rights agency that arranges deals with streaming services for many independent labels, found that the income guarantee offered to them by one streaming service amounted to *six times* what they would ultimately have earned through royalties alone. Deals like that are how the bulk of the payments that streaming services send out end up staying in the pockets of the labels.

Getting People to Care
About Your Work

IT USED TO be that one of the most substantial advertisements for a book was the book itself. Sitting on a shelf at a bookstore, or in a spinner rack at an airport magazine stand or a grocery store, the book and its cover were positioned to draw people in. It makes sense: after all, everyone who wanted books knew that the way to get them was to go to bookstores. That meant that everyone in a bookstore was interested in books. The customers delivered themselves to the products, and shopped through a finite set of items until they located the ones that suited them.

People still go to bookstores, and video stores, and music stores. But all these sectors are in sharp decline, and in many cases the most significant channel for creative work is now the Internet. Customers don't necessarily deliver themselves to "stores"—virtual or physical— and when they do, the titles on offer are rarely the neatly curated, finite, and browsable selections that once dominated.

The shelves, instead, are nearly infinite. Browsing has been augmented by search algorithms and automated recommendation systems. And the number of ways for customers to discover new work has exploded.

Word of mouth has always been a creator's best friend. Recommendations from personally trusted sources were a surefire way to sell products. When I worked in a bookstore, one of the most reliable indicators of an imminent sale was two friends entering the store together, and one of them picking up a book and handing it to the other with the words "Oh, you've *got* to read this; you'll *love* it."

The Internet is practically made of word of mouth. Telecommunications has always been bigger than entertainment. U.S.

telecommunications businesses—companies that let people talk to other people—brought in $750 billion in 2011. The U.S. entertainment sector, in 2012, brought in $480 billion.

Content isn't king. Conversation is.

Content Isn't King

PEOPLE USE THE Internet to communicate with others about everything that matters to them. There isn't really such a thing as a "channel for content" and a "channel for talking about content"—on the Internet, the same video-hosting services that offer the best place to post (potentially infringing) clips of commercial video are *also* the best places to post legitimate videos of people talking about the commercial videos they love. Twitter isn't just a place for posting links to potentially infringing downloads of copyrighted material—it's also the place where people post links to long discussions of the copyrighted material they love, recommending it to other people.

What's more, these services are also the place where people talk about *everything else*. A conversational, social medium where you're allowed to talk only about music—and not your life and times—will be stilted and unconversational. Even the most dedicated, topical forums on highly technical subjects—cancer therapy, high-energy physics— inevitably coexist with nearby forums for "chatter" about pop culture, personal thoughts, and idle chitchat.

Indeed, this is practically the Internet's origin story: the U.S. government created a military and scientific network for information sharing, and its users promptly started a *Star Trek* discussion forum. Tim Berners-Lee created the World Wide Web for sharing high-energy physics papers, and its users promptly started posting pictures of their cats, their failed cake-baking adventures, and the titanic snowfall that had just been dumped outside their lab windows. And then they, too, started arguing about *Star Trek*.

Sociable conversation is the inevitable product of socializing. Sociable conversation is the way that human beings establish trusted relationships among themselves. And sociable conversation among trusted parties is where invaluable recommendations for creative works arise ("Did you see X on TV last night?"). On the Internet, every medium is first a medium for social conversation and secondarily a specialized forum for some other purpose. Anything we do to make it more expensive and difficult to create, host, and link to new mediums also makes it harder to have the social conversations that turn random strangers into paying customers for creative work.

How Do I Get People to Pay Me?

CONVERTING AUDIENCE APPRECIATION into something you can use to pay for your kids' braces is the crucial task for any artist. There are shelves full of MBA-program case studies and countless online seminars from various successes and would-be successes enumerating theories about how to do it. Nearly everyone who tries their techniques will fail—that's just the way of it—but there are several approaches that still make sense in the Internet era. The following overview is a guide to how the basic strategies of creative business have worked in the past, and how they've been successfully adapted to today's Internet-era reality.

1. YOU CAN SELL A PHYSICAL COPY OF YOUR ART

You (or a publisher, a label, or a studio) can still make a DVD, a CD, a book, a USB stick, a print, a sculpture, or some other tangible embodiment of your creation. You can charge money to own these things, or to rent them. It's not always easy to keep physical stuff from being taken without payment—a cat burglar may be waiting to steal your painting out of your studio right now!—but it's infinitely easier than keeping people from making digital copies and distributing them without your permission. You're probably familiar with this business model. It's an oldie, but a goodie.

2. YOU CAN SELL ADS

Before the net, getting an advertiser was hard, but the high barrier to entry kept the prices *charged to advertisers* high. So a newspaper, if successful, could make a lot of money from the space it gave over to ads. Today, newspapers—and other traditional media—make a lot less that way. But that's not because the process of finding advertisements to run against your content has gotten harder.

The Internet has made it easier than ever to get money for displaying ads around your stuff. But easiness isn't all there is to it. Now that there are lots of places for advertisers to go, all of which are competing for the same ads, the price an advertiser needs to pay has gone down. This has been offset somewhat by the rise in companies looking to advertise—it's gotten easier to do that, too!—but as any newspaper publisher will tell you, the increase in demand hasn't kept up with the increase in supply.

The fundamental problem that "traditional media" is having is that its business was structured around expensive, resource-intensive undertakings and paying large dividends to investors. Newspapers bought purpose-built buildings in central New York and Tokyo; radio networks took over enormous towers next door to them; record labels built multimillion-dollar studios and employed titanic numbers of administrators, talent scouts, managers, and so forth.

The net makes it possible to do things more cheaply. For one thing, the actual production costs for media have fallen drastically. It's not *easy* to do professional typesetting, but if you know how to do it, you can make it happen with the computer under your arm, and you can pocket the difference between the cost of a computer you already own and the cost of a

Costs and creativity

"Cost disease" is an economic concept that is vital to understanding the relationship between labor costs and the arts. First described by William J. Baumol in the 1960s, cost disease describes the way that technology seems to drive up the cost of "services." (In econo-jargon, performing music and writing books are both "services.")

Technology generally reduces the amount of labor needed to make physical things. Every year, automation drives down the number of human hours needed to assemble a car; and thus, every year, for the average buyer, cars tend to get more affordable, as their labor costs decrease.

Services are much harder to automate. Teaching a kid how to read, examining a patient, performing a sonata, or cooking a hamburger all take approximately the same amount of time now as they did a century ago. And since the people who do these jobs all expect to be able to buy cars and other manufactured goods, their wages can't be (fairly) discounted just because their "product" isn't getting cheaper as quickly as manufactured goods are. More often, in fact, their wages go up.

This is, incidentally, one of the key reasons that education and health care command ever-larger slices of developed countries' GDP—the machines and buildings get cheaper, but the people who operate them are a fixed cost.

This same dynamic impacts the arts. Whether you're writing a novel, performing a song, or painting a painting, there are some costs—both in time and money—that can't be reduced.

However, there are many parts of a creative living that *have* gotten vastly cheaper thanks to technology. Twenty years ago, I was the CIO of a successful documentary-film-production house. Our edit suite cost $250,000, and didn't even produce production-quality output—after films were edited on it, they had to be re-cut on a multimillion dollar system that we rented access to. Our cameras—digital betas—cost several times more than modern SLR and Red digital cameras, and produced footage that was nowhere near the same quality.

Today, you could replicate our whole setup for much less than ten thousand dollars. And that's not all: these days, when your crew goes on location, it can book its own plane tickets—no travel-agent fees—shop around for customs processing, save big money with Airbnb and hotel discounters, and so on.

The time filmmakers spend writing their scripts and recording their interviews and editing down their footage costs just as much as it ever did. But every other cost has gone down. These are the *capital* costs—the costs that you'd typically borrow or raise funds to cover. The stuff that you need time for has stayed fixed, but time is something you can provide on your own, without begging patrons or investors for help. Meanwhile, the cost of the stuff that you have to sell your soul and vision for—the cameras, the plane tickets, the hotel rooms and edit suites—has plummeted, and there is no bottom in sight.

Yes, it takes an orchestra the same number of hours today to perform a Mozart symphony as it did in the eighteenth century, but all the other costs of delivering that symphony to the greatest number of listeners are vastly cheaper today than they've ever been. Rehearsing and performing are the things that the orchestra can do for itself, while everything else is the stuff they have to sell their soul for. Cost disease hasn't reduced the cost of performing, but it has given musicians much more control over their destiny than they've ever had before.

huge typesetting shop full of specialized equipment that cost a million dollars twenty years ago. The hyperexpensive shots that George Lucas stuck into *Star Wars* in 1977 can now be rendered cheerfully and without complaint by a used PC that your local high school is throwing away. That doesn't mean you, personally, know how to make that PC produce something as cool or lucrative as Lucas did, of course—but if you can, you have a lot more options than Lucas did back then for making money from it, because the cost is so low.

So if analog dollars have turned into digital dimes (as the critics of ad-supported media have it), it's worth remembering that it's possible to run a business that gets the same amount of advertising as its forebears at a fraction of the price. You can still profit from a much smaller income, as long as you have much smaller expenses, too.

3. YOU CAN SELL SWAG

There's no question that the market for certain embodiments of art has declined. For example, I've got no interest in ever acquiring a CD again—a CD isn't an asset, it's a liability. When I get a CD, I have to rip the disc, make sure the song titles and other metadata were correctly transferred, and then figure out how to get rid of the CD itself in a way that is both legal and environmentally responsible. (If you give

xkcd

Randall Munroe is a funny guy and a moderately talented illustrator. Trained as a physicist, he started posting his humorous stick-figure doodles to his website, xkcd.com. (He says the name doesn't mean anything—it was just a four-letter .com domain he managed to snag before they were all snapped up by speculators.) Eventually, he began to produce a regular comic strip. He uses Creative Commons licenses on his comics that permit their unlimited noncommercial sharing. His strips have a heavy science/technology bent (the comic's strapline is "A webcomic of romance, sarcasm, math, and language"), and you see them posted on office doors in every university's math, science, and computer science departments. They're also liberally posted around research institutes like CERN, home of the Large Hadron Collider, and the Wellcome Trust Sanger Institute, one of the homes of the Human Genome Project.

But Randall can't eat the adulation of nerds. Instead, he sells swag. A *lot* of swag. Randy lives with his wife in a house whose living room has been converted into a ball pit. He and his friends sit in the ball pit and play video games all day. Three times a week, Randall draws and posts an xkcd comic. His other major chore is depositing checks from the company that sells his T-shirts and posters. He's living the dream.

What's more, the more famous—the more *copied*— Randall's comics get, the more money he makes. He doesn't need to control or reduce copies. Good thing, because he knows he can't.

it away after ripping it, it's probably not legal; if you throw it away after ripping it, it's *definitely* not environmentally responsible.) This is one reason that "piracy" statistics from the music industry are so misleading: they imply that every downloaded song is a lost sale, and that every lost sale is a lost sale of the whole album, not just the single. But if piracy vanished tomorrow, people like me wouldn't start buying CDs again. We don't want those liabilities in our lives. The best you could hope for is that a small fraction of today's downloaders would become iTunes or Amazon MP3 customers, which is a lot less commercially exciting than turning them into buyers of $17.99 CDs. There're those analog dollars/ digital dimes again.

But there are plenty of high-margin physical goods that don't simply reproduce an artwork, but rather represent some *affinity* for it. T-shirts, posters, and every manner of tchotchke and gimcrack and gewgaw can be sold to fans who want a chance to express their identity by publicly displaying their taste in media. Some artists have also turned largely obsolete formats like CDs into swag by packaging them in elaborate, beautiful enclosures. For example, David Byrne and Brian Eno's *Everything That Happens Will Happen Today* album was released in a limited edition that included a CD, a DVD, and a miniature diorama train set with light-activated sound effects. (I have one and treasure it.)

What's more, this is the era of on-demand swag. Increasingly, T-shirts and other items can be made in very small batches, even one at a time, which allows creators to try out a lot of designs without committing a lot of capital, or ending up with an attic full of unsold merch.

4. YOU CAN SELL COMMISSIONS

Creators who are well loved for their work often attract commissions from companies and individuals. Rich people are infamous for commissioning fashion designers to produce one-of-a-kind outfits for important events; restaurateurs commission murals; advertising agencies commission commercial jingles; stock-art agencies commission pictures. I've written several commissioned science-fiction stories. Some were for textbooks that needed short fiction to accompany a technical passage; a few were for high-tech companies that were doing future-product planning and wanted fiction to spark their engineering discussion; one commission was sold off directly in exchange for an ad that accompanied the story on publication.

One way to think about commissioned work is that it represents the price of adjusting your creative priorities. For example, many free/open-source software creators work on

Madonna and the concert promoter

In 2007, the singer Madonna walked out on Warner Bros., her record label of twenty-four years, and signed a $120 million deal with Live Nation, a concert promoter, allowing it to take up distribution of her records and produce a string of incredibly lucrative mega-tours. Concert promoters don't particularly care if recorded music is paid for or copied freely, just so long as it's popular, because the more popular an artist is, the more the concert tickets are worth. In 2011, Madonna and Live Nation partnered with a label, a Universal division called Interscope, to release a new album, her first since 2008; the tour that followed far outearned the record itself.

programs and features just because they like the idea of them. But, having made a name for themselves as expert, high-quality software developers, these people attract commercial commissions from companies that have a need for a specific feature or program. So the programmer takes a break from working on her own priorities and turns her attention to someone else's, and pockets a commission in exchange.

There have always been creative agencies that specialize in this sort of commission, but increasingly artists can avail themselves of services like deviantART and other portfolio sites as a free or cheap way to hang out a shingle for potential clients.

5. YOU CAN SELL TICKETS

If you create things that can be performed, there is an ancient and honorable way of making money from them: you can perform them in a room with a door that closes. You station someone imposing at that door, and instruct that person to ask anyone who wants to come in to buy a ticket. If they don't buy a ticket, they don't get past the door.

2013 was history's top-grossing year for live music performance. This is a good time to be a live performer, not least because of all the ways the Internet supplies for getting the word out about shows. Now, not everyone is cut out to perform—I do a lot of touring and speaking, and it's hard work. I miss my family, I don't get enough sleep, and it's hard to keep up with my writing. But if it were easy, everyone would be doing it.

Back when records and radio were invented, many musicians hated them. "I'm a live performer," they said. "I do something as old and as holy as the first story told in front of the first fire. How dare you tell me that I am to be a mere clerk, doomed to sit in a back room while you *technicians* make my work available to my public?" That kind of mind-set leaves you driving taxis and flipping burgers.

Eighty years later, the spiritual descendants of the musicians who succeeded as recording artists have a different complaint. "I don't *perform*. I'm not a trained monkey. I am a white-collar worker. I labor indoors. When my work is done, I slide it under the door and some bourgeois man of commerce takes it out to the world. What right have you to tell me that in order to earn a living, I must become an itinerant *jongleur* who capers for others' amusement?"

In both cases, the refuseniks misunderstood how technologically determined their income was. Art is art, whether you make it with a computer-based mixing board or by banging two rocks together. But *industry* is all about technology. There was once a thriving lamplighting industry—people were paid to walk the streets with long, flaming poles that were used to light the wicks on public streetlamps at dusk. Those jobs were obliterated by the electric light. The tragedy of the lamplighters who failed to find another trade is real, and should never be dismissed, but that doesn't make electric lights immoral. When it comes to business, technology giveth and technology taketh away.

6. YOU CAN ASK FOR DONATIONS

This may be the oldest business model there is for entertainers and artists: asking people around you to voluntarily give you money so you can go on making more of the stuff they already see and hear. Again, the Internet acts as a force multiplier here—you can ask more people, in more places, and accept their donations in more ways.

The Humble Indie Bundle is a wildly successful "pay-what-you-like" distributor of video games. Several times a year, the Humble Bundle people announce a new "bundle"—five to seven video games, all sold together. The games are shipped without any digital locks, and will play on Macs, PCs, and GNU/Linux computers. Customers are invited to name a fair price for the bundle, and are shown how much other customers are giving on average. To spur their competitive natures, buyers are given stats broken down by operating system: "You're on a Windows

PC, and Windows users are giving an average of $40. Mac users are giving an average of $43, and Linux users are giving $48." Customers are also allowed to go back later and donate more money, if they feel like they underpaid. And even though people can pay anything, from zero dollars to thousands, the bundles typically make in excess of a million dollars each. Each customer specifies how much of their payment should go to a few charities nominated by the Humble project, and the rest of the money is divided up among the games' producers—a hundred thousand to three hundred thousand dollars each. Users also have the option to add a "tip" to Humble itself, to pay for the administration of future bundles, and that brings in enough to keep the lights on.

With my help, Humble has now branched out into e-books and comics, and further refined its sales pitch. Recent bundles have included an embeddable "widget" that each creator can put on his or her web page, as well as unique author-specific links. That way, Humble knows which artist's work got you interested in the bundle, and it's able to use that intelligence to prick at your competitive nature: "You're a fan of Author X. You and your fellow fans are giving an average of $54, which is well below the fans of the next-most popular author, Y, who are giving an average of $67." They'll soon do the same thing with automatically detected cities: "You're coming in from San Francisco. San Franciscans are ranked eighty-seventh in global payers; below Oslo, ranked eighty-sixth, where the average payment is $58."

Humble isn't the only innovative collector of donations. Kickstarter uses "crowdfunding" to raise money for creators—people solicit funds to complete a project, and make a pitch (text and video) explaining why donors should trust them to use the money wisely. Then they specify premiums and gifts to be given to exceptional donors—give ten dollars and I'll send you a postcard with a custom sketch; a hundred dollars gets you a custom portrait; ten thousand dollars gets you an original comic book starring you and your friends. Kickstarter has also been used as an effective means of collecting preorders before a production run: Give me fifteen dollars, and I'll send you a book. Once I have enough fifteen-dollar commitments in hand, I can have the books printed and ship them out.

Finally, you can always just stick a payment button on your website. Creators have varying luck with this strategy, but maybe you've got the right combination of audience, material, and pitch. One caveat: PayPal, the most popular payment processor online, has a well-deserved reputation as a high-handed, obstreperous bureaucracy that arbitrarily freezes its customers' accounts, often without recourse. Hardly a week goes by without some high-profile company, individual, or charity going public with their tale of PayPal woe. I haven't experienced this myself (yet), but I make a point of moving my funds from PayPal to a real bank account as quickly as possible, never leaving more than a few dollars in my PayPal account at any given time.

None of this precludes chasing the established patronage systems—arts-council and NEA grants, institutional money, private funding—but as patronage has gone mass and global, the chances of you finding someone or some group of people willing to fund your vision have radically expanded, as well.

Molly Crabapple and the distributed patrons

Molly Crabapple is a talented painter who works in large-form canvases. She's the sort of painter who, in another era, might have worked through a gallery and an agent, trying to please wealthy people who might part with astronomical sums for her works. A few painters have made it that way, but most have failed.

Crabapple, though, turns out to have a real flair for crowdfunding. In 2011, she launched her "Week in Hell" project, where she locked herself in a hotel room for five days, papering the walls with poster-paper and then decorating every inch of the paper with illustration. The event was meant to commemorate her twenty-eighth birthday, and she sought to raise $4,500 on Kickstarter from fans who got to watch her draw on a live video feed and received pieces of illustrated paper. She raised $25,805.

In 2012, she sought $30,000 from her fans to rent a New York City storefront and paint nine giant paintings inspired by the Occupy Wall Street movement.

Within a week, she had $55,000. By the time the project ended, she had $64,799 (including the $8,000 per painting she took in for the seven canvases she sold as part of the highest-level Kickstarter reward). Like many painters through history, Crabapple relies on patrons to pay her bills—but her patrons number in the thousands, and she needn't worry about the caprice or high-handedness of a few fat cats as she paints her way into history.

Does This Mean You Should Ditch Your Investor and Go Indie?

No.

Well, in some cases, yes. If you've given a lot of thought to your industry, have a good feel for what's going on, and are prepared to do a lot of work that isn't related in any particular way to "creating"—that is instead best characterized as accounting, administration, marketing, PR, bookkeeping, and sales—then you're less likely to fail on your own than you are to fail with an industry partner.

Media companies aren't stupid. They're big, and they're sometimes slow, and their priorities aren't the same as those of the creators whose work they market. They want to make money across a portfolio of works, and they need to make money in a way that doesn't damage their critical business relationships. They're not going to freeze out major distributors just because the distributors want digital locks. Their sales forces will be distracted when their lead titles are entering the pipeline, and they may ignore your creation if it happens to be slotted for the same time. They need to make enough money to support their infrastructure— their offices, their equipment, their staff—even though much of that infrastructure won't be used for putting your works into the hands of audiences. And they've got to have enough left over at the end of the day to keep their shareholders happy, which means that their business practices will be designed to ensure that shareholders get money out of the business before creators do. After all, they can always get new creators to fill their pipeline—but investors are strictly non-commodity, and are the true first-class citizens of a media firm. This isn't a truth limited to media businesses—it applies to every for-profit company.

So it's not insane to think that you might be able to do a better job of getting your work into the hands of your audience than a publisher would.

But keep in mind the advantages that media companies bring. They have economies of scale, so they can produce and finish your work more cheaply than you might—for example, they have boilerplate legal agreements for illustrators and designers who work on your marketing materials, as well as longstanding working relationships with large groups of artists with proven track records for producing effective materials. They can find the right artist, commission an illustration, sign the agreement, and pay that artist far more cheaply than you can, all other things being equal.

They have an efficient, tested workflow. They can take a raw work and turn it into a finished work for sale in the places where customers expect to buy it, without having to learn anything new or solve any problems. You will have to learn new things and solve a lot of problems to do the same thing, and it will almost always cost you more and take you longer.

They have capital. In the event that there are production costs to be borne before you can sell your work, media companies can act as investors. They sometimes put money into your hands before you finish the work, too, though always with many strings attached.

But then again, media companies don't have monopolies on these things. You might know people who can act as your sales force, or your PR team, or your marketing firm; you might know designers, or editors, or distributors. If you've devoted a lot of study to the state of your market, and if you have evolved a theory of how to do business within it, and if you want to take time away from creating to be a publisher, studio, or label, then this can be a very lucrative and exciting project. Just remember that the mere fact that entertainment giants are giant, slow-moving, and remorseless does not mean that they aren't also the way to make the most money while reaching the largest audience with your work.

Love

THERE'S ONE VITAL consideration that needs to be touched on when talking about choosing between going indie or signing with a media company: love.

Boing Boing started out as a zine—an indie magazine—complete with indie orthography: the name was written *bOING bOING*. The founders, Mark Frauenfelder and Carla Sinclair, were a married couple who published the zine as a labor of love—but because it was a printed, physical good, it got more expensive for them as it got more popular. Recognizing this, Mark and Carla charged for copies. I used to sell it in the late 1980s, when I was working in bookstores. It was one of my favorite publications.

bOING bOING did well for a "little" magazine, and found a national distributor that went belly-up shortly thereafter. Many independent zines, including *bOING bOING*, died with it. When book and magazine distributors go bankrupt, they often take down publishers, too, because the publishers' inventory is sold off to pay the banks and other creditors. So publishers lose both their merchandise and all the money that their distributors owe them, which is usually all it takes to bankrupt them, too.

Mark and Carla took up other projects after that, and many years later Mark found himself covering the launch of a new service called Blogger for a magazine called the *Industry Standard*. He still had the boingboing.net domain, and so he "revived" the magazine as a blog, posting a few items a week for a year, for the enjoyment of himself and a few friends.

But in early 2001, Mark broke a major story. He dug up the patent drawings for a secret device created by famed inventor Dean Kamen that had been code-named "Ginger." Everyone was abuzz about what

Ginger might be. (It turned out to be the Segway scooter, which pretty much failed to revolutionize the world the way its investors had claimed it world.) CNN featured *Boing Boing*'s home page on the air, and Mark's readership shot up exponentially overnight.

Mark was heading out on holiday the next day, and he was worried that if those new readers came back, they'd give up on the site because nothing new had been posted. So he asked me if I'd "guest edit" the site while he was away. When he came back, he invited me to make my position permanent. Over the years, we've added two more friends as editors, Xeni Jardin and David Pescovitz, and the four of us have built the site up to a very large readership.

In 2003, that readership started to break the bank. We saw our bandwidth bills jump from about fifty dollars a month to over a thousand dollars a month, and *Boing Boing* stopped looking like a fun hobby and started looking like a drain on our bank balances. So we contacted John Battelle, the guy for whom we'd all written at *Wired* magazine, and who had also founded the *Industry Standard* and assigned Mark to cover the Blogger launch. John had been playing with the idea of founding an ad brokerage for sites like ours, and he was eager to prototype his business idea by seeing if he could sell ads for *Boing Boing*. He did, and went on to hire a sales force that represented us and many other sites. He called his new company Federated Media, and today it remains our major ad partner in what has become a multimillion-dollar business.

But from the beginning, *Boing Boing* (and even *bOING bOING*) has been a labor of love. I write it because it's my major way of thinking through the stuff that matters to me. It's a repository of all my thoughts and inspirations, my public notebook and my soapbox. The other editors feel much the same way. Many times over the years, we've turned down large sums of money from would-be buyers because they went against that ethos.

If all the money went out of *Boing Boing* tomorrow, it would have virtually no effect on my writing there. *Boing Boing* exists because it makes its owners very happy. It also happens to make us money. But we did it before it made us a penny, for years, and without any particularly urgent desire to commercialize it.

Creators usually start doing what they do for love. There are creators who say, "If I couldn't earn a living making art, I'd do something else," but it's hard to take this seriously. Pursuing an arts career is not the move of a rational mercenary. Almost everyone who sets out to earn a living from the arts will fail. Almost all those who succeed will make a very poor living indeed. Entering the arts because you want to get rich is like buying lottery tickets because you want to get rich. Though, of course, someone always wins the lottery.

You might find that, whatever its other benefits or drawbacks, being your own investor/publisher/label/studio is incredibly satisfying. If you love doing it, you should do it. You might not make any money, but you probably won't make any money any other way.

Funny story: the *Industry Standard* spiked the story on blogging on the grounds that it was just a passing fad.

The New Intermediaries

HOWEVER YOU DECIDE to handle the independence question, unless you're also up for being an ISP, a payment processor, a retailer, a wholesaler, and a marketing company, you're going to have to sell your works through one or more intermediaries. Intermediaries are vital to creative business, making it easier for ever-larger pools of creators to get paid for their work. If the only way to get your videos out there is to host them yourself, then the pool of successful video creators will be limited to those people who can make great movies *and* great video-hosting tools. Thankfully, we have YouTube (and Vimeo, and the Internet Archive, and VODO, and Netflix…).

However, when competition is scarce among intermediaries—when there are only a few ways to get your payments processed or your e-books sold—the companies that control those channels will turn them into bottlenecks, and will use their power to extract as much money as they can from the creators who depend on them.

For as long as there have been middlemen who sit between a supplier and a customer, there have been debates over how much responsibility the middleman owed to both, and to the wider society. Intermediaries make for easy targets when you're trying to solve hard business problems. If you're concerned about counterfeit handbags entering the country, you could solve your problem very cheaply just by making the nation's port authorities legally liable for any fakes that land onshore. Then *they'd* have to spend the money checking every container that arrives from the Far East. This would cost billions, and delay shipments of every other kind of good sent to America by container ship (which is basically everything), but it wouldn't cost *you* anything. And if the ports refused to cooperate, you could go to the rail companies, or the trucking companies, or the toll-booth operators, all the way down to the credit-card processors and the banks.

Economists call this "externalizing," which sounds a bit like a term from family therapy. A company "externalizes" its costs by causing some other business (or the government) to pick up the tab. The classic externalized cost is pollution. A company dumps sludge into the water supply instead of paying to process it, and everyone who relies on that water has to pay to de-sludge it before they can drink it. Another feature of externalizing: it usually ends up costing more, overall, than dealing with the cost internally. It's a lot cheaper to process sludge at the source than to get sludge out of water, but everyone pays a chunk of the latter cost, while the polluter alone bears the former one. So for the polluter, externalizing makes sense.

Technology creates new intermediaries, and changes what it's possible to require of them. The advent of steam engines created the railroads. The advent of telegraphy created the telegraph companies. And over and over again, everyone from law enforcement to business owners to crusading moralists have sought to make the intermediaries of their day do something to help their causes. In response, the prevailing legal approach to intermediaries has been to treat them as "common carriers."

A common carrier is an intermediary that "carries" everyone who can pay the price of admission, regardless of who they are, where they're coming from, where they're going, and why they're out and about. In exchange, common carriers are absolved of any liability for the problems created by their carriage. A ferryman who accepts every cart that's bound for the market is a common carrier. A phone company that wires up anyone's house, regardless of race, creed, color, or reason, is a common carrier. And a freight company that accepts cargo from all comers is a common carrier.

Being a common carrier requires that you take on certain responsibilities to wider society. Your ferry will be subject to more-stringent safety regulations than a private boat. Your phone company will have to cooperate with the police when they show up with a wiretap warrant.

The ferryman loses his common-carrier status as soon as he starts to pick and choose from among the people on their way to market. If he carries only Farmer Sam's eggs and refuses to take anyone else's, he is now a part of Farmer Sam's operations; if Farmer Sam poisons the town with rotten eggs, the ferryman might find himself hauled up for questioning alongside him. Similarly, if your business spins off a division that runs and manages phone lines to all your satellite offices, that division will be a party to any businesswide fraud you commit. If your train line hauls pig iron only for the Rockefellers and not the Carnegies, then you will have some responsibility to ensure that you're not helping Rockefeller swindle the tax man.

The Internet is full of intermediaries. There's the ISP that provides your connection to the Internet, which might be a telecommunications company like AT&T, or a business or other institution, like a library, café, employer, school, or hotel. Often, there might be one or more ISPs "upstream" of that ISP.

Then there are online service providers (OSPs), which might provide you with a message board, a blogging system, a video-hosting service, a file locker, an instant-messaging system, or an email box.

There are software companies (or foundations, or authors) that make peer-to-peer file-sharing clients, or mobile apps, or the operating systems these programs run on.

There are the app stores and download sites that provide access to software.

There are other kinds of stores that sell access to other sorts of digital products, or that sell you physical goods that are sent to you by mail.

There are game servers that allow multiple players to play together. There are voice-over-IP companies that allow for in-game chat or video conferencing.

There are ad brokerages that supply advertising to ad-supported sites and programs. There are domain registrars, which sell Internet domains like www.thepiratebay.se.

There are payment processors like PayPal, MasterCard, and AmEx, which help get money from customers to suppliers.

And in the age of "cloud computing," there are companies that host massive arrays of raw computing power, supplying bandwidth, storage, and processing.

All of these intermediaries can play a critical role in any commercial creative venture. In fact, it's nearly impossible to imagine doing anything of consequence without recourse to one or more of them. Even a "solitary" project like writing a novel will involve an operating-system vendor, a word-processing vendor, an email-hosting company (to communicate with your editor and agent), an ISP (so you can access the email), an email-software vendor, and probably a few online file lockers for previewing cover art, exchanging unwieldy electronic galleys, and the like. And whether you go indie or through one of the big five publishers, you'll likely be expected to use some combination of blogs, Twitter, Facebook, and YouTube to promote your work. You might "visit" libraries or schools by Skype. If you go out on tour, you'll want to use something like Upcoming.org to get the word out, and you might use a payment processor to sell tickets to some of your events. And, of course, the book itself will be sold through Amazon, BN.com, and the online shops run by independent stores.

On top of that, there's the world of e-books, whose intermediaries include specialist file-conversion houses that turn InDesign print layout files into ePubs, Mobipocket files, PDFs, HTML 5, and a million other potential formats, and the companies that sell those e-books, from the aforementioned Amazon and Barnes & Noble to Google Books and all its independent booksellers. And there are the e-book hardware vendors, who may or may not have the power to stop you from loading a "bad" book onto your device.

So what's the difference between these newfangled intermediaries and all the ones that came before them? Scale. Unimaginable scale.

A really top-notch cable operator might carry two hundred to five hundred TV channels, each one airing ten to twenty-four hours of programming a day. Assuming your cable operator had two hundred channels, that's a minimum of two thousand hours of video a day. As of March 2014, YouTube was adding that much content *every twenty minutes*. A cable operator might conceivably be required to ensure that a copyright lawyer has inspected all the video that enters its distribution channel, but if we apply that same requirement to YouTube, we would end up exhausting the entire lifespan of every copyright lawyer who ever lived and still not make a dent.

Net neutrality

2014 was the year that "net neutrality" sprang into the public consciousness, and not a moment too soon. At its heart, net neutrality is the idea that ISPs should deliver the bits we ask for as quickly as they can get them. ISPs, on the other hand, are petitioning for the right to give favorable treatment to some kinds of Internet data. For example, if YouTube bribes your ISP for "fast lane" access to its customers, you'll have great, speedy access to YouTube—and all its competition will be jittery and sucky.

ISPs say that anyone who objects to this arrangement wants something for nothing. Why should YouTube's competitors have the same access to AT&T's subscribers if YouTube is willing to pay extra for "premium" access?

The reality is that ISPs are trying to get paid three times for the same service through network discrimination. You pay for your home broadband connection. The website you're visiting is also paying for its connection. The ISP then wants to treat you as its hostage and ransom you to that website.

A useful parallel here would be voice service. Imagine that your corner pizza shop makes the best pizzas in town. It's so good that it's got ten phone lines with phone-answerers standing by to take your order. But it hasn't paid the phone company for "premium" access, because it couldn't outbid Domino's. So when you call it, about half the time you'll get a busy signal—even if it has non-busy lines, with people standing by to answer them—and you'll have to call back. Meanwhile, every call to Domino's gets put through right away.

You paid for your phone, and in return you want the phone company to connect you to all the phones you dial. Joe's Corner Pizza paid for its phone lines, and in return it wants to be able to talk to anyone who calls. But AT&T wants to pick the winners in the pizza wars, handing over the ability to reliably receive orders to the deepest-pocketed pizza shop on the block.

Joe's doesn't want something for nothing, and neither do you—you both just want the phone company to do its job.

The ISPs say they can't afford to upgrade America's pitiful, trailing-edge garbage networks unless they can extort these ransoms from Internet users and services. This is such an obvious lie that it's amazing anyone takes it seriously. According to the carriers' own investor filings, their major profit centers are already their Internet businesses, and yet they devote only the barest minimum to capital investments in maintenance and upgrades.

The carriers are creatures of regulation, companies that couldn't exist without some of the most valuable government handouts this side of the defense industry. That's because the cost of negotiating for every yard of road and sidewalk they have to dig up to lay their wires would cost trillions. Instead, governments use regulations to give them access to "rights of way" that make this cost bearable.

A robust regulator could solve the problem of net neutrality at the stroke of a pen. The chairman of the Federal Communications Commission could say, "Look, we gave you trillions in rights-of-way savings so that the public could get the network it needs. If you want to run your network for maximum profit without considering the public interest, get your damned wires out of our dirt and start negotiating your own rights of way. You have six weeks, and after that, we'll pay you the scrappage rates for the copper. Don't worry, we'll find any number of companies that'll be happy to take this giant subsidy and give the public back its due."

But the FCC is not a robust regulator. As of 2014, the chairman of the FCC is Tom Wheeler, whose last job was *serving as the top lobbyist for the cable industry*. As John Oliver quipped, "This is the equivalent of needing a babysitter and hiring a dingo."

Intermediary Liability

CABLE OPERATORS AREN'T "common carriers." That means they have to contend with liability.

In copyright land, strict liability means that if you participate in a copyright infringement, you are liable for it, even if you were acting in good faith and taking reasonable steps to stay on the right side of the law. Cable operators aren't quite held to this standard, though their liability does come close. So if you're a cable operator and one of your five hundred channels carries a live newscast that runs a copyrighted photo, not only can the photographer sue the broadcaster and the person holding the photo—they can also take a stab at suing *you*. It doesn't matter if the newscast is from a reputable station that has broadcast for years without ever straying into infringing territory. It doesn't matter that you couldn't possibly prescreen a live news show. You still might end up on the hook.

Now, obviously this is an impossible situation. Cable operators have deep pockets, and there is a lot of programming in the big cable multiplex. It's a juicy target for nuisance copyright suits. So how do operators manage this? Through a chain of indemnity. Every channel in the multiplex signs a legal document indemnifying the carrier for infringement, and promising to carry enough insurance to pay off in the event of a real suit. Every show on every channel signs a legal document indemnifying the channel in the same way. Every supplier to every show does the same, but for the show. Ultimately, that means that a nuisance suit against an operator will quickly run into a bunch of well-paid, angry insurance company lawyers who'll eat it for breakfast.

But the quid pro quo from the insurers is a punishing "compliance" regime in all areas of artistic production. In order to get your movie or TV show insured, you have to demonstrate to an insurance officer that you have gone well beyond what copyright law requires of you, so that there's no doubt those insurance lawyers would win any suit that arose. This means, for example, that documentary filmmakers need to get permission from companies whose logos are shown on the T-shirts of real-life rioters in real-world situations (which means that if a company doesn't want to have its brand associated with those events, it can effectively censor the shot just by withholding permission). Copyright law doesn't require this kind of permission, but the insurer doesn't want to leave any room for doubt.

So now let's scale things up again.

Imagine what strict liability would mean for, say, Twitter. At present, Twitter fields around six thousand tweets per *second*. Any one of these could infringe copyright. And digital copyright infringement carries punishing liability: up to $150,000 in damages and $250,000 in fines. If *one hundredth of 1 percent* of Twitter's tweets infringed copyright, the potential hourly liability approaches a billion dollars. And that's before you've paid the lawyers.

Notice and Takedown

STRICT LIABILITY JUST doesn't scale.

"Common carrier" law *does* scale, but the entertainment industry (or the publishers and investors behind it, at least) has never been a fan of the notion. After all, common carriers have virtually no duties to industry, apart from dealing fairly and evenly with all comers. If Warner Music Group called up AT&T to complain that someone in Des Moines was holding up their phone to a boom box and transmitting copyrighted music to a confederate in Arroyo Alto, Texas (population 363), AT&T could shrug its mighty shoulders and turn away.

The Goldilocks middle ground here is Notice and Takedown (NaTD)—the *other* part of the WIPO Copyright Treaty. Notice and Takedown works pretty much like it sounds: if you think a file hosted on my server is infringing on your copyright, you tell me (Notice) and I have to remove it (Takedown). As long as I do so, I'm considered a common carrier—no matter what happens in the ensuing legal battle, I'm not liable for it. If I don't take it down, I'm in strict-liability land: if you win a judgment against the guy who's using my server to infringe on your copyright, you can sue me for damages.

If there's one word that comes up consistently in discussions of NaTD, it's "balance." As in, "This is a way to balance the rights of creators and investors with the business realities of online media."

But in practice, the actual participants in NaTD pretty much all hate it. Media companies say that searching every online service for copies of their material is an impossible task: if you've got a million hours of TV programming in your vaults, and you have to check every video-hosting service, BitTorrent tracker, and file locker for any unauthorized clips, *and* pay a lawyer to send a takedown notice in every case, you're going to spend a hell of a lot of money without making much of a dent in the amount of material available. The hosting companies, meanwhile, counter that it's no fun for them to get hundreds of thousands of takedown notices per day. They claim they want to do right by their customers, and make sure those takedown requests they receive pass the giggle test, which means that NaTD costs them money, too.

To top it all off, there's a vast cohort of Internet trolls, bullies, crooks, and scammers sending fake takedown notices to get rid of stuff that embarrasses them. Under Notice and Takedown rules, it becomes trivial to silence one's political enemies, or people whom you simply disagree with. The Chilling Effects Clearinghouse has cataloged hundreds of thousands of takedown notices, many of them utterly spurious. Examples of takedown abuse include:

- Police departments whose officers have been recorded committing illegal acts, claiming copyright on the videos of these acts.
- Diebold using takedown notices to suppress a memo detailing its deliberate sale of flawed voting machines in violation of federal election law.
- The Church of Scientology using takedown to attack opponents who published church documents that, to many, shed critical light on the religion.

There are almost never penalties for abusing the takedown process; it's the measure of first resort for rich and powerful people and companies who are threatened by online disclosures of corruption and misdeeds. Hosting companies say they get floods of spurious takedown requests from media giants who don't even bother to download the files they're claiming infringe on their copyrights. Collectively, the hosts spend millions on trained staff to evaluate copyright claims, but no matter how much they spend, there just aren't enough eyeballs to examine all the beefs that everyone on the planet might have with their customers. And why should owning some computers in a data center somewhere require them to become de facto arbitrators of entertainment law, anyway?

Very few of us have the skill to create our own Twitter-like service, or YouTube-grade video-hosting site. But many of us have a lot to say on this sort of service. Any hurdle that the law presents in founding and operating such a service is a tax on all the people who might use that service to express themselves. Likewise, anything that makes it easy to remove material from the Internet makes it easy to commit acts of petty censorship.

Remember, *creators* hate this regime, too. If you're unfortunate enough to have your material incorrectly flagged as violating someone else's copyrights, you find yourself in a topsy-turvy world where the presumption of innocence is nowhere to be seen. Instead, you're left trying to convince an administrator at an ISP or web-hosting company that you're on the right side of a law they don't understand very well. What's more, you'll probably be talking to someone who doesn't have to understand it—their job depends on their bosses staying out of legal jeopardy, not on making the right call about your material.

So What's Next?

THE ENTERTAINMENT INDUSTRY has exceptional lobbying prowess, and it's extremely adept at locating sympathetic lawmaking forums in which to pitch its case. Globally, copyright is often interwoven into trade agreements. For example, every country that joins the World Trade Organization also has to sign something called the Agreement on Trade-Related Aspects of Intellectual Property Rights (the TRIPS Agreement, a precursor to the WCT), which in turn requires them to agree to the Berne Convention. So if your country wants to join the WTO, it has to sign the Berne Convention, too.

In the last few years, the Office of the U.S. Trade Representative has been working through closed-door "plurilateral" negotiations to create additional copyright treaties with its major trading partners. Starting in 2008, the Anti-Counterfeiting Trade Agreement (ACTA) was negotiated in secrecy between ten countries and the EU. (The EU negotiator refused to disclose treaty drafts to the European Parliament, and objections by its member countries ultimately led to the EU's withdrawal from the treaty, at the end of 2012.) Here in the U.S., 2011 saw the introduction of the Stop Online Piracy Act (SOPA) and its senate equivalent, the Protect-IP Act (PIPA). PIPA and SOPA sparked a global storm of Internet-driven protests, and resulted in millions of calls and emails to the U.S. Congress and Senate and, ultimately, the withdrawal of the legislation.

Millions of people intuitively understand that the Internet is more than a glorified video-on-demand service or a more perfect pornography-retrieval system or a plain old great way to shop. They understand that the Internet is the nervous system of the information age, and that laws that treat it without due regard are a threat to the very idea of a fair and free society. Unprecedented numbers of people have risen up against ACTA and SOPA.

But the ideas behind those proposals aren't dead. They live on in treaties like the Trans-Pacific Partnership (TPP), which has been undergoing formal negotiations since 2010, and the Transatlantic Free Trade Agreement (TAFTA), which has been under negotiation since 2013. Both are being brokered outside the public eye.

Though SOPA, PIPA, ACTA, the TPP, the WCT, and their ilk differ in their specifics, they share certain broad themes that represent the legislative agenda for the entertainment lobby. And if you wanted to sum up that agenda in a single sentence, it would be this:

More intermediary liability, with fewer checks and balances.

More Intermediary Liability,
Fewer Checks and Balances

THE INTERNET HAS a *lot* of intermediaries. At present, treaties like the WCT focus on controlling one class of middlemen: web hosts, be they cloud-computing providers, file lockers, video hosts, or social-media services. Their liability is limited as long as they're willing to respond to a takedown notice. Beyond that, intermediaries aren't required to police their customers' uploads.

What the entertainment lobby wants is to expand *which* intermediaries are liable for their role in an infringement, and *what* responsibilities intermediaries have when it comes to proactively policing their services. So who are they targeting?

Hosting providers and other online service providers: These are the companies that currently operate under the Notice and Takedown rules. Under agreements like SOPA, PIPA, and the TPP, online service providers from Facebook to YouTube to Twitter to Tumblr would become more like cable companies—they'd be responsible for ensuring that nothing infringing is posted to their services in the first place, rather than merely being required to respond to notices from rights-holders about infringement.

Linkers: Under SOPA and PIPA rules, any site that hosts a link to *another* site that links to infringing content would be a party to that infringement. That means that if, say, Facebook published a user's link to a website that itself contained links to infringing content, Facebook could be sued.

ISPs: Currently, the companies that connect you to the Internet have very little liability for your actions. If you use your home DSL or your office network to send an infringing file from A to B, the people who hooked you up to the Internet are not a party to your actions, as long as they satisfy some minimal formal requirements. But under a series of proposals with names like "Three Strikes," "Notice and Termination," and "Graduated Response," ISPs would have the legal obligation to disconnect customers who generate copyright-infringement complaints. France's Hadopi law and New Zealand's Copyright Amendment Act 2011 were the pioneers here—in both countries, users were threatened with total disconnection from the Internet for a prolonged period on the basis of unproven accusations of infringement. (Thus far, punishment under the laws has been limited to fines.)

Search engines: A search engine's job is to "crawl" over the Internet, make an index of everything it finds, and then serve up lists of links in response to users' queries. So if you type "dog" into a Google search box, Google sends you a list of all the pages it knows about that are relevant to "dog," in ranked order. Under proposals like SOPA and PIPA, search engines would be required to ensure that they serve up only non-infringing links.

Payment processors and ad networks: At present, payment processors (like American Express, PayPal, MasterCard, and Amazon Payments) are not responsible for ensuring that the money they take from Alice to give to Bob is being transferred for a legitimate purpose. Under proposals like SOPA and PIPA, though, payment processors would be required to freeze the accounts of customers who are accused of abetting copyright infringement. Likewise, ad brokers (companies that match ads with websites, and pay the websites for displaying their ads) would have to respond to accusations of infringement by cutting off the ad supply of websites that have been accused.

Registrars: Registrars are companies like VeriSign and Hover that sell domain names to companies, organizations, and individuals. Under proposals like SOPA and PIPA, registrars would have to cooperate in the seizure of domains from customers who stand accused of infringement.

DNS providers: Most ISPs (and specialist companies like openDNS) provide "domain name services" to domain owners. This involves running a computer that converts domain names ("craphound.com") to numeric IP addresses (204.11.50.137), a service that is critical to the way the Internet works. Under proposals like SOPA and PIPA, DNS providers would have to respond to accusations of infringement by suspending such services for the accused domains.

Proxies and circumvention tools: Many companies and nonprofit organizations maintain "proxy" servers, which can be used to increase the security and robustness of one's Internet connections. For example, millions of workers use their companies' proxy servers to securely access the wider Internet—they connect to a local network, which establishes a secure connection to the company's proxy, which then mediates all further traffic. At the same time, proxy tools like the Onion Router, which started as a project of the U.S. Naval Research Laboratory, help millions of people to evade national firewalls, such as those used in Syria and China, in order to access the free press and disguise their identities. Under proposals like SOPA and PIPA, producing a tool like this would be banned, and the researchers, companies, and programmers who maintain these tools and services would be breaking the law.

Technology designers: Computers are copying machines, as is the

Internet. We use metaphors like "reading" a file or "loading" a page, but what happens in each of those instances is that files are *copied*, often several times and in several kinds of buffers and caches, before you see or hear them. Most of the post-Internet copyright rules treat "ephemeral" copies as acceptable—you don't need a copyright license to make a separate copy of a document in a network card's buffer, another in RAM, another in the video card's buffer, another in a drive cache, and so on. Under policies like the TPP, though, ephemeral copies made in buffers could require a separate copyright license. This would have the effect of giving rightsholder groups indirect oversight of the designs of networks, PCs, operating systems, applications, mobile phones, and subcomponents like video cards, hard drives, and network cards.

Many of the intermediaries mentioned above can already be pressed into service by rightsholders, but only through the intervention of courts or arbitrating bodies. If you want to get a domain name revoked, you can make a complaint under the auspices of the World Intellectual Property Organization's Uniform Domain Name Dispute Resolution Policy and argue your case. If you want to get a hosting service to proactively filter all the uploads it receives to check for your copyrights, you can sue them to try to make this happen. (Viacom tried this in 2007, suing YouTube for one billion dollars in damages it claimed stemmed from illegal broadcasting of its copyrighted material; YouTube—which subsequently became part of Google—tightened its copyright-protection policies even as it won several judgments in the courts. The case was ultimately settled, without money changing hands, in March of 2014.)

The post-SOPA initiatives now under way take some measures to "streamline" these processes. In the case of SOPA and PIPA, rightsholders wouldn't have needed to go to court to prove infringement before getting action from web hosts, ISPs, DNS companies, payment processors, ad brokers, or domain registrars—instead, they could just supply lists of the businesses and individuals they'd like to see action against, and the intermediaries would have been bound to act. Lobbyists are still pushing for such powers, via legislative channels and otherwise.

Meanwhile, the massive consolidation of the ISP industry has created another danger: "voluntary" agreements between ISPs and other industries, allowing for arbitrary censorship and surveillance without any kind of rules, oversight, or court orders. The Obama administration's IP czars—who rotate into the job and then back out into the entertainment industry—have brokered several of these deals.

What these policies have in common is an expectation that intermediaries should be policing interactions between their users. The companies involved are private actors, but they're encouraged to act more like gatekeepers than service providers. If you want your videos posted to YouTube, you now have to "audition" them for YouTube's copyright inspectors to confirm they pass muster. If you want to get money out of your PayPal account, you might have to prove to PayPal that you haven't been infringing on any copyrights. And if they suspect otherwise, they might just disconnect you and keep your money.

Disorganized Channels Are
Good for Creators

SO WHAT WILL putting all this pressure on intermediaries do to creators? It's a fair bet that any regulation that makes it more expensive,
and more difficult, to be an intermediary will reduce the number of
intermediaries. YouTube was founded by three guys and some venture
capital; if the next YouTube requires that you spend a thousand dollars
on lawyers for every dollar you spend on hard drives and bandwidth,
there won't be any more YouTubes. And that means that the existing
YouTube-like services will stabilize, consolidate, and settle on the least
competitive terms they can all live with.

The fewer channels there are, the worse the deal for creators will
be. Any choke point between the creator and the audience will turn
into a tollbooth, where someone will charge whatever the market will
bear for the privilege of facilitating the buying and selling of creative
work. Economists call this "transfer pricing"—all of a sudden, profits
are captured at this choke point, rather than at any previous or suc-
cessive stage.

Right now, there are a *lot* of distribution channels that creators
can use to reach audiences—if you don't like YouTube, there's Hulu,
Vimeo, Vodo, and a hundred others. If you don't like Amazon, there's
BN.com, Lulu, Smashwords, BookBaby, and many other platforms.
If you don't want to sell your video games through Walmart, there's
Steam, the Humble Indie Bundle, and many other venues.

In the old days, investors—labels, publishers, studios—could sew
those channels up. There were only so many sales calls that Walmart's
buyers would entertain, so if you wanted to get a book or CD onto its
shelves, you had to sign up with one of the companies that had a good

working relationship with those buyers. You could always go indie and try to make it on your own, but the vast majority of customers for your creations would have no access to your product. Getting a zine or a CD onto the shelves of a local store meant that local audiences might discover it, but it took a big, organized middleman to get the same CD into the hands of a broad base of potential customers.

Today, your iTunes or Amazon MP3 track is exactly the same number of clicks away from a listener as anything that Universal Music Group puts out. You still have to figure out how to get potential customers interested in your music, but assuming you manage to do that, your customers can buy your music instantaneously, in the same places they buy music from the big guys. There are far too many distribution channels for the big three labels or the big five publishers or the big six studios to be able to dominate them. These companies may have very efficient processes for getting their products into these channels, and excellent promotions, but if you don't want to do business with them, or if they don't want to do business with you, you can still sell your creative work to anyone who wants to buy it from you.

What's more, the distribution channels that exist today run the gamut from established giants to spunky start-ups, and there's a lot of volatility in the sector. Yesterday's giant is today's bankrupt loser. Every company with a distribution business knows this, and they vie with one another to offer the best terms possible to creators. If your publisher gives you an 8 percent retail royalty and sells your book for seventeen dollars with a 40 percent wholesale discount, Amazon (which discounts new releases by 20 percent) can step in as publisher, offer the same book for ten dollars with a 30 percent royalty, and still make the same amount on each sale. So they *really* want you to ditch your publisher, and they'll try to tempt you with a rather good deal.

This means that a publisher that wants to keep its authors will have to offer better deals to them in turn. When there's a small number of buyers for creative works and a large number of creators, creators get a bad deal. When the number of buyers goes up, the deal gets better.

That's why it isn't in creators' interests for the operating costs of new distribution systems to go way, way up.

Freedom Can Be Expensive,
but Censorship Costs Us the World

CREATORS ARE OFTEN asked to add their names and credibility to letters to Congress, public campaigns, and other efforts to increase intermediary liability. It's understandable that musicians are upset that, for example, so little of the money that streaming services like Pandora pay out ends up with them, but when they locate the problem with streaming services—instead of with the labels, who pocket almost all those payments—they make it harder for artists to get paid. If the big labels continue to constitute the only legitimate voice in the music copyright debate—if no new venues for music distribution can thrive—then the deal for musicians won't ever get better.

If you're a creator, your ability to earn a living is directly tied to two things:

1. Your negotiating leverage with the companies and people who control the channels between you and your audience.
2. The ease with which your work can be discovered by potential customers.

Everything we do to increase intermediary liability decreases creators' leverage and makes it harder for audiences to help each other find new things to love.

Having leverage and finding an audience isn't a guarantee that you'll earn anything, of course. But you can't earn a single penny until you have those two things.

New creators' groups like the Authors Alliance seek to add more moderate, thoughtful voices to the debate. These groups don't give the

Internet giants a free pass, but they understand that just because they're mad at Google and their publishers are mad at Google, it doesn't mean that they want the same things as their publishers.

It's up to creators everywhere to engage with their colleagues about the ways that expanded liability for intermediaries drive us all into the old-media companies' corrals, where they get to make the rules, pick the winners, and run the show.

Information Doesn't Want to Be Free, People Do

B ACK IN 1984, Stewart Brand—founder of the *Whole Earth Catalog*— had a public conversation with Apple cofounder Steve Wozniak at the first Hackers Conference. There, Brand uttered a few dozen famous words:

"On the one hand, information wants to be expensive, because it's so valuable. The right information in the right place just changes your life. On the other hand, information wants to be free, because the cost of getting it out is getting lower and lower all the time. So you have these two fighting against each other."

This is quite a good Zen koan, a compact meditation on the duality of information as an economic good and as a technical matter. But thirty years on, the phrase has gone from a useful way of provoking discussion about the philosophy of the information society to a trite slogan that obscures more than it illuminates.

It's time to kill it.

The "desires" of information are totally irrelevant to the destiny of the Internet, the creative industries, or equitable society. Information is an abstraction, and it doesn't "want" anything.

Information doesn't want to be free—people do.

That's my third law. That's what this book is about. What people want from computers and the Internet. What we stand to gain, and what we stand to lose.

What the Copyfight Is About

I'VE SPENT THE past ten years in the "copyfight," the often bizarre, multi-faceted political struggle over the destiny of the Internet and copyright. It's taken me to street marches, courtrooms, legislative bodies, treaty-making UN agencies, and special commissions, and I've met thousands and thousands of people from all walks of life who want a free and open Internet.

These people have devoted their lives to fighting against special protection for digital locks. They're fighting, too, against increased intermediary liability on the Internet.

Not one of them has done this because of what "information" wants. Rather, the fight is, and always has been, about people.

Copyfighters aren't in the streets because they worry about the fortunes of titanic technology companies or titanic entertainment companies. In truth, they're not even particularly exercised about the fortunes of people who want to earn a living in the creative arts. Ultimately, that stuff is of interest to a tiny fraction of the world's population.

But the open Internet—and efforts to close it—affects every one of us.

Beyond art

A common tactic in discussions about the Internet as a free-speech medium is to discount Internet discourse as inherently trivial. Who cares about blurry kitten pictures, illiterate YouTube trolling, and Facebook posts about what your toddler said on the way to day care? Do we really want to trade all the pleasure and economic activity generated by the entertainment industry for *that*?

The usual rebuttal is to point out all the "worthy" ways that we communicate online: the scholarly discussions, the terminally ill comforting one another, the distance education that lifts poor and excluded people out of their limited straits, the dissidents who post videos of secret police murdering street protesters.

All that stuff is important, but when it comes to interpersonal communications, trivial should be enough.

The reason nearly everything we put on the Internet seems "trivial" is because, seen in isolation, nearly everything we say and do *is* trivial. There is nothing of particular moment in the conversations I have with my wife over the breakfast table. There is nothing earthshaking in the stories I tell my daughter when we walk to daycare in the morning. This doesn't mean that it's sane, right, or even possible to regulate these interactions.

Taken together, these "meaningless" interactions make up nearly the whole of our lives. They are the invisible threads that bind us to our friends and families. When I am away from my family, it's these moments that I miss. Our social intercourse is built on subtext as much as it is on text—when you ask your wife how she slept last night, you aren't really interested in her sleep. You're interested in her knowing that you care about her. When you ask after a friend's kids, you don't really care about their potty-training progress—you and your friend are reinforcing your bond of mutual care.

If that's not enough reason to defend the trivial, consider this: the momentous arises only from the trivial.

When we rally around a friend with cancer, or celebrate the extraordinary achievements of a friend who does well, or commiserate over the death of a loved one, we do so only because we have an underlying layer of trivial interaction that makes our connection to these people meaningful. Weddings are a big deal, but every wedding is preceded by a long period of small, individually unimportant interactions. Without these "unimportant" moments, there would be no marriages.

The copyright wars are about *all* the things we care about on the Internet, and increasingly that encompasses just about everything in our lives. Every time we make it harder to put text, audio, video, and files on the web, we limit the kinds of things that people can say and do in this new, networked public sphere. Right now, the Internet teems with innumerable personal communications: Facebook updates, Twitter tweets, blog posts and comments. People from all walks of life have found a new way to converse on subjects trivial and grave. Some of it is serious, some of it is silly, but very little of it would exist if every time someone wanted to open her mouths online, she had to pay the cost of having her speech vetted by a process intended to catch copyright infringement.

What proportion of YouTube is "copyrighted material"?

Oh sure, the entertainment industry says, *YouTube could be used for anything, but in practice it's used to rip us off.* But what this ignores is that YouTube holds a *lot* of video. A hundred hours of new material are uploaded every minute. Multiply every entry in the Internet Movie Database (which also includes broadcast TV shows) by an average of ninety minutes per program (low for movies, high for TV shows) and you get about twenty-eight days' worth of YouTube uploads. Simple math tells us that only a fraction of YouTube can possibly consist of directly infringing clips, uploaded for the sole purpose of helping people see and hear otherwise expensive things for free.

Google doesn't publish statistics about who watches what on YouTube, but network analysis performed in other online media would suggest that the material on YouTube follows a "long tail" curve, in which most of the page views accrue to a small fraction of super-popular material, and the rest go to material that is so obscure as to be invisible in traditional media. That would be kitten videos, heartfelt outpourings by adolescents, political rants, independently produced music and video, and other examples from the whole wide world of human expression. Even if you assume that the popular stuff on YouTube is just ripped-off commercial material (it certainly isn't, but this is just for the sake of argument), then that still leaves a huge quantity of YouTube content that has nothing to do with the entertainment industry.

Two Kinds of Regulation

THERE'S NOTHING WRONG with the idea of a big, high-stakes industry having some legally enforceable rules—hell, I'd be delighted if the finance industry could get some meaningful regulatory oversight! But if banking regulations started to creep into everyday life, even their staunchest defenders would be dismayed.

Most of us are comfortable with the idea that banks that loan lots of money to one another should have to document their actions carefully and disclose their liabilities to regulators on demand. On the other hand, we would all find it a little tedious if these same rules were applied every time you decided to pick up the tab for lunch with a friend.

Industrial regulations should apply to industries, not individuals, families, and private groups. Every industrial regulation includes some kind of test to help you figure out whether your activity falls into the regulated realm or can be considered a private matter. We say that loans below a certain dollar value are excluded from banking disclosures. But if the U.S. was hit by hyperinflation and a sandwich suddenly cost a million dollars, you'd expect the loan threshold to be revised to match. If it wasn't, you'd get into a situation where loaning a friend enough money for lunch was suddenly subject to SEC oversight.

The Internet has given us hyperinflation for copying. Copyright's test for industrial activity—are you making or handling a copy?—is no longer a good way to sort entertainment-industry transactions from personal, cultural, private activity. Insisting that normal people, doing normal things, should be able to navigate a system designed for a big, sophisticated industry is a fool's errand.

As a system of regulation for the entertainment industry, individual copyright laws can be good or bad. I like the part of copyright that says my publisher can't print this book without getting my permission, because I won't give them that permission unless they pay me. I don't like the part of copyright that says that I, the author, can't authorize you to break digital locks that are put on my works by an intermediary.

But disagreeing with some rules doesn't mean you disagree with rules altogether. Wanting a different copyright isn't the same as not wanting copyright at all.

Anti-Tank Mines and Land Mines

BEYOND QUIBBLES OVER which copyright rules we should have, there is the even more pressing question of whom those rules should apply to.

The rule of thumb that copyright uses to figure out if you're part of the copyright industry is whether you are making copies. This made perfect sense in the last century. Anyone who was pressing a record probably had a million-dollar record factory. Anyone printing a book probably had a printing press, a bunch of skilled printers to keep it running, and a building to house it all.

Equating copying with industrial activity made sense when copying was hard. The legal scholar James Boyle describes this as designing copyright the way you design an anti-tank mine—anti-tank mines are designed to detonate only when you drive over them with a multi-ton tank. Anything lighter than that—a civilian car, or a civilian on foot— is ignored by the tank's detonation mechanism. Anti-tank mines don't always work perfectly, but when they do, you can (in theory) put them all over the place, even in playgrounds, and the only time they'll blow up is when someone shows up in a tank.

The problem is that over time, computers have made copying exponentially easier and cheaper. It's as though we planted a bunch of anti-tank mines around the playground, and fifty years later new manufacturing techniques have put safe, innocent, actual-size toy tanks within reach of every ten-year-old. Suddenly, the anti-tank mine becomes an anti-personnel mine, and a system that was supposed to interact only with instruments of war starts going off indiscriminately, with a bunch of non-combatants inside the blast radius.

Put it this way: it makes perfect sense that the lawyers at Universal Studios should have to talk to the lawyers at Warner Bros. when Universal decides to build a Harry Potter ride. But when a twelve-year-old wants to post her Harry Potter fan fiction or the Harry Potter draw-ings she made in art class on the Internet, it makes *no* sense for her to negotiate with Warner's lawyers. She can't afford to pay a lawyer to advise her, and even if she could, no one at Warner's would find it worth their while to talk to her, anyway.

And moreover, there's nothing wrong or new with making Harry Potter fanfic or drawings. Kids have been doing this forever; every suc-cessful artist learns her trade by copying the things she admires. That's why the streets of Florence have a copy of Michelangelo's *David* on every corner—Florentine sculptors learn to sculpt by copying the acknowl-edged all-time city-wide champion sculptor.

Technically, copyright may have prohibited things like this before (although *David* is safely in the public domain). But before the Internet, it was much more difficult for a rightsholder to discover that an offense was taking place, and there was very little pressure on intermedi-aries to police copyright on the rightsholders' behalf. No one asked the companies that sold school notebooks to ensure that fanfic was never scribbled in their pages. No one asked art teachers to ensure their students were staying on the right side of copyright in their fig-ure-drawing classes.

But all this changes in an era of Internet-scale intermediaries, net-worked communities, and automated Notice and Takedown procedures. Instagram or Twitter becomes the preferred way for kids to share their drawings with one another; Fanfic.net becomes the preferred place for fanfic authors to share their work with one another. Technically, the companies providing these services are "making money off copyright infringement," but no more than the mall food court near the local high school makes a few bucks off the students who gather there to show off their infringing art while eating lunch.

In truth, there has always been too much infringing material out there to expect it to all be policed by regulators. And as we've seen, that volume of unpoliced content has only grown with the advent of the Internet. The difference is that the regulation is becoming automated. The copyright-bots that YouTube now employs to evaluate its content don't make any distinction between industrial copyright infringement and what I think of as "cultural activity." Your fan video is caught in the remorseless, relentless, and fully automated enforcement systems set up by rightsholders and Google, and can be taken down by a process that is entirely untouched by human hands. No one working for the intermediary or the entertainment company has the time or money to look at every automatic match and make sure they're not being unreasonable—instead, they have an automated anti-tank system, and they can't figure out how to stop it being triggered by kids. Lacking any way to improve the trigger, they just leave it where it is, and catch tanks and toys alike.

Technically, copyright has always prohibited you from making your own copies of record albums and your own prints of feature films. My grandparents were legally enjoined from copying the 78 rpm records they collected. But for them—and nearly everyone else of their generation—a rule saying "You may not copy records" was about as superfluous as a rule that said "You may not carve your name into the face of the moon with an enormous laser." The main reason the music fans of the 1950s couldn't copy music was that they lacked access to a record press. The law was entirely beside the point.

Laws that are beside the point can say all kinds of silly things, and the silly things will be beside the point, too. The reality is that as soon as the capacity to copy music (and, later, video) for personal reasons reached the average person, the world's courts and legislators started creating a web of laws and rules that legalized this activity. They recognized that there was a difference between a music bootlegger setting up an illegal press to run off competing copies and an individual who makes a mixtape for a friend or records something off the TV to watch later.

The Internet era has conjured forth mountains of nonsense about the death of copyright. Reformers have claimed that copyright is dead because the Internet makes it impossible to control who copies what; copyright supporters have said that the Internet itself must be contained, to head off that grim fate.

This is rubbish.

It's impossible to control who loans a friend lunch money, but that doesn't mean financial regulation is dead. It just means that financial regulation has to limit itself to the kinds of transactions that take place on an industrial scale, among industrial players. A copyright regulation that is sophisticated enough to handle all the nuanced business questions that the industry encounters can *never* be simple enough for the majority of Internet users to understand, much less obey. And a copyright that is simple enough for a twelve-year-old Harry Potter fan to understand will *never* be sophisticated enough to regulate the interactions of billion-dollar entertainment conglomerates and their suppliers and vendors.

The ease of copying in the modern world has nothing to do with whether Warner Bros. can sue Universal for creating unlicensed Harry Potter theme parks. It has nothing to do with whether authors can sue publishers who print their books without securing the rights. It has nothing to do with whether movie studios can sue online stores that sell their movies without authorization, or cinemas that screen them without paying for them.

Copyright is alive and well—*as an industrial regulation*. Copyright as a means of regulating cultural activities among private individuals isn't dead, because *it's never been alive*.

Is the entertainment industry dying?

The Internet era has been attended by a parade of new copyright laws and proposals, each one justified by the supposed near death of the entertainment industry. The stats behind these proclamations of imminent doom are not particularly reliable. The U.S. Government Accountability Office, as neutral a body of economic statisticians as you're likely to find, concluded that it is "difficult, if not impossible, to quantify the net effect of counterfeiting and piracy on the economy as a whole." Of the piracy-impact studies most often cited by the entertainment lobby, the GAO said, "[They] cannot be substantiated or traced back to an underlying data source or methodology."

It is probably true that *some* companies' and *some* creators' fortunes are affected by piracy, and by the expansion of the pool of entertainment choices that audiences can make when they spend their money. But in January 2012, a thoroughly documented Computer and Communications Industry Association study called "The Sky Is Rising" found that more creators are creating more media, and that larger audiences are spending more money, than ever before. There's now more music, more movies, more video, and more books being created than at any other time in history, and they're bringing in a larger pool of money than ever.

However, the money is a lot less "lumpy" than it has been in the past, especially in the recent past. Since the creation of communications technology, entertainment has been a winner-takes-all kind of business, with the most successful creators outperforming the least successful by huge margins. Think of how, before the record player, local singers could always find work playing the popular songs of the day. But after the invention of records, local audiences could buy recordings of the most popular performers performing the most popular songs, and the local second choices were no longer in demand.

After seventy or eighty years of this, the pendulum is swinging back again. The dramatically lowered barriers to producing and distributing media and the new, easy tools for discovering new artists mean that audiences are fragmenting—buying more overall, but less from any one publisher. The number of self-published works on Amazon is orders of magnitude larger than the number of books published by the traditional Big Five. There are thousands of times more hours of video being produced by indie creators for direct sale or ad revenue (or both) on YouTube and Vimeo than the big media companies could ever hope to generate. Kickstarter, meanwhile, has totally changed the world of games publishing, and webcomics offer homes to far more cartoonists than ever found a place in syndicated newspaper funny pages.

The condemnations of this new age of accessibility often focus on the "low quality" of the material that the new players produce. It isn't as glibly written, as skillfully edited, or as expertly recorded as the material we're used to from the traditional entertainment industry.

But "quality" isn't the same as "production values." As the old saying goes, "You can't polish a turd." Every zero-star movie from a major studio is "high quality" in that it has been produced by skilled craftspeople who ensured that the turd was as shiny as possible.

A badly shot video of someone you care about has merits that can never be matched by a skillful video of a stranger. Heartfelt personal messages from people you love—or people you identify with—have virtues that can't be matched by scripted banter. Desperate outpourings smuggled out of dictatorships and tyrannies have an immediacy and authenticity that dry newspaper reportage can't match.

We don't have to choose, of course. The reports of the entertainment industry's imminent demise are vastly exaggerated. But as we debate the future of communications and free speech, let's not discount as worthless, just because they lack artifice, the personal, homely, handcrafted messages that speak straight to our hearts.

Hollywood versus Google

The future of the Internet should not be a fight over whether Google (or Apple or Microsoft) gets to be in charge or whether Hollywood gets to be in charge.

Left to their own devices, Big Tech and Big Content are perfectly capable of coming up with a position that keeps both "sides" happy at the expense of everyone else.

Take YouTube, for example: now that it's impossible to start an effective competitor to YouTube—because you'd need an army of copyright lawyers and a database of all the copyrighted video ever made in order to run your own Content ID system—YouTube and the major labels have cut deals to run a music-streaming service that perfectly suits the labels' needs.

But some indie labels have not opted in to this system. In mid-2014, Google announced that any indie label that didn't make its music available for the streaming service would no longer be allowed to use YouTube to promote its catalog. And because running a YouTube-like service is so legally fraught, there's nowhere else for the indies to turn for the same vital promotional value.

If you're worried about tech companies converging on any given business model—advertising, surveillance, "native" advertising in which companies get to design art—the best way to fight back is to be sure that a few companies from tech or entertainment don't end up running the show.

There's nothing inherent to being a tech startup or an entertainment startup that makes you treat artists like shit. It's about scale: once a company is big enough that it can boss everyone around, it does, and that bossing is only ever to the benefit of that company and its shareholders.

Who's Talking?

IN THE LAST chapter, I talked about how existing systems like Notice and Takedown, for all their flaws, at least attempt to provide "balance" through some sort of dispute-resolution mechanism. Under Notice and Takedown, users whose online material has been removed can file formal objections explaining why they deserve to have it reinstated. SOPA and PIPA included provisions allowing users to unfreeze financial accounts and reinstate websites by filing legal papers with the appropriate intermediaries.

If we were talking about competing businesses, these procedures would make a certain amount of sense. (Not much, though.) If you can figure out how to start a company, run a payroll, and file your corporate taxes, you can presumably ask a lawyer to help you sort out a bit of trouble with your website.

But everyday life doesn't have a business model. The messages that teenagers send to one another may be key to their emotional development and education, but that doesn't mean they have the organizational wherewithal to maneuver through dispute-resolution systems designed for corporations with conflicting interests. And this goes double for people who don't speak the language that the dispute-resolution system uses—dissidents in far-off lands, immigrants, and people who go cross-eyed at the sight of legalese.

Think how many vital social activities have moved online, in whole or in part. Is there a politician who could get elected without the web, now? How many of us are there who found a job through the Internet, and now rely on it to help us perform our duties?

In 1991, George Holliday's video of Los Angeles Police Department officers delivering a savage beating to Rodney King was a shocking anomaly, something that most of the world had never seen. Now police abuses are routinely recorded and posted online, where they become too big for traditional media to ignore. The same goes for camera-phone footage of ballot-stuffing in Belarus, official corruption in India, Molotov cocktails in Kiev's Maidan, and parental abuse all over the world.

We live in a highly mobile world, where increased migration has created vast diaspora populations. My family was part of an earlier diaspora: my grandparents fled the USSR after World War II and came to Canada, where they all but lost touch with their relatives back home. I saw my great-grandmother (who lived in Leningrad) only a small handful of times before she died. But today's diasporas remain tightly bound together, able to route around high long-distance tariffs with voice-over-IP calling, and they enjoy services that could exist only with the Internet, including video calling and video- and photo-sharing. This is particularly important for connecting preliterate children or illiterate adults with one another. My six-year-old daughter delights in her videoconferences with her grandparents and cousins in Toronto from our flat in London.

The point is: if you regulate the Internet as though it was a glorified system for delivering cable TV or listening to music, you'll miss the main event. There are huge populations who rely on the net to live as full-fledged members of society—their actions are the most important things happening on the Internet, and whatever else we do to help one industry or another weather the digital shift, we must protect these uses, too.

In 2009, the UK government's Champion for Digital Inclusion commissioned a PricewaterhouseCoopers study on the impact of Internet access on the lives of some of the poorest, most vulnerable people in the country. It compared outcomes for families living in the hardest-hit subsidized housing projects in northern England, where the collapse of heavy industry has sent unemployment skyrocketing. The researchers controlled for obvious variables—education and background and so on—and looked at similar families with and without Internet access.

What they found was surprising, even for Internet advocates. Families with Internet access had a higher quality of life by practically every measure. They enjoyed better health and nutrition and more social mobility; their children got better grades and were more likely to go on to post-secondary education; the parents had better jobs and were able to save more money from those jobs; and the whole household was better informed about politics and more engaged with civic projects. A similar study conducted in the U.S. in 2013 drew similar conclusions. In short, net access moves the needle for social policy.

The consequences of losing Internet access, on the other hand, are getting graver and graver. Today, virtually everything we do involves the Internet, from grocery shopping to planning a potluck dinner. Tomorrow, virtually everything may well *require* it. And yet our policies about Internet access continue to lag behind that reality. It's a human-rights catastrophe in the offing.

Is Internet access really a human right?

The UN, the EU, Finland, and many other governmental entities describe Internet access as a human right. But not everyone agrees. Vint Cerf, the distinguished computer scientist whose work was critical to the very invention of the Internet, published a 2012 op-ed in the *New York Times* saying that access to the Internet wasn't a human right in itself, but merely a conduit for *delivering* human rights.

I understand where he's coming from, but respectfully disagree. To understand why, consider the notion of a "free press." In the USA, the First Amendment guarantees a free press and bars government censorship. So imagine that the government sent police to a newspaper's printing press and seized all the presses on political grounds, and then said that the First Amendment was intact because the paper's owners could continue to "publish" by photocopier, or over the Internet, or by shouting their message through a bullhorn. Variations on this scenario have played out in U.S. jurisprudence, and the judges who've been called upon to consider the question of whether censorship of some media is acceptable, so long as some other medium remains available, have decided that it is not.

It's clear to me today that the full spectrum of activities that compose free speech, a free press, and freedom of assembly cannot be realized without Internet access. Yes, it's still (barely) possible to organize a mass demonstration without the net, by running a phone tree, or sticking posters up on phone poles, and getting the permits from one of the few remaining human clerks at city hall. But the hurdles faced by someone whom the state has deprived of network access are titanic, relative to the easy path her networked colleagues enjoy. A "free press" means more than "You are free to hand-write your message on scraps of paper and hand them to people"—it requires access to the full range of press technologies, and that includes the Internet.

Censorship Doesn't Solve Problems

IN SPITE OF all the research we now have on the good effects of Internet access, many countries are still preoccupied with keeping their citizens off of it. Internet blackouts—disconnection—have been a favorite tactic everywhere from Mubarak-era Egypt to junta-led Burma to China's Xinjiang province during Uighur uprisings. Most people who think about the Internet in China immediately think of the country's Great Firewall, a massive censorship system that traps every single Internet request and decides, on a URL-by-URL basis, whether the person making the request should be allowed to load that file. Even liberal democracies with strong free-speech traditions have resorted to national censorwalls to screen out objectionable or illegal content. These initiatives usually begin with a stated desire to screen out child pornography. Child pornography is one of those content categories that no one defends—it's something that we all (rightly) recoil from. As the father of a six-year-old, the thought of its production turns my stomach.

So we propose censorwalls. They aren't particularly effective. Even the proponents of censorwalls admit that they're easy to get around, and so they refuse to publish lists of what's being censored and what isn't. If censorwalls worked, it wouldn't matter if you knew which websites they blocked, because you wouldn't be able to see them, due to the effective censorwall. But since censorwalls don't work, publishing a list of what's being blocked is the same thing as publishing a kind of yellow pages for finding child pornography on the net.

Once you agree that censorwalls are necessary, and that censorwalls don't work well, you inevitably end up concluding that the censors' blacklist has to be a secret. That is to say, censorwalls can't be transparent and accountable. And indeed, that's how all the censorwalls in the world work, from the blocklist running on your corporate network to the one at your kids' school, up through the child-porn filters operated by national governments and their arm's-length agencies in Canada, the UK, Sweden, and Australia, all the way up to the censorwalls used by undemocratic governments in Bahrain, Syria, China, Uzbekistan, and Saudi Arabia. That's a problem—there's plenty of stuff that governments and their agents would like to suppress, including extreme political beliefs, pornography that doesn't involve children, and material that provides counseling on contentious issues, from reproductive health to assisted suicide to drug legalization advocacy.

Something else about child pornography: it's pretty clear-cut. While there are some weird edge-cases that reasonable people might disagree about—the nude art-photos that Lewis Carroll took of small children, say—they are the exception, not the rule. When I type the words "child porn" and you read the words "child porn," we're probably thinking of the same thing. There is broad social consensus on what is—and isn't—child pornography.

But if child porn is relatively unambiguous, copyright infringement—another censorwall target, in many places—is very muddy. There's no way to tell, merely by looking at a file, whether it has been put online with the authorization of the rightsholder. Even if you're certain that the material is there without authorization, it's still a hard problem to determine whether it's infringing, or whether it falls into one of the many exceptions to copyright—such as fair use, in the USA, or fair dealing, in the rest of the world. Meanwhile, there is little social consensus on what constitutes reasonable copyright—many people believe that downloading a movie they own on DVD or a song they own on CD is fair game, though the law would almost certainly find this to be illegal.

Furthermore, a copyright censorwall's blocklist will have to be secret for the same reason that a child-porn censorwall's list is secret. And since there's a lot more disagreement about what constitutes copyright infringement than there is about what constitutes child porn, all the ambiguity-related problems—overbroad filtering—will be a lot worse. Once we allow for some authority to secretly, unaccountably block information, we'd be foolish to expect anything but overbroad, abusive censorship.

Even worse is the fact that today, censorship is inextricably bound up with surveillance. Back in the 1930s, when James Joyce's *Ulysses* was banned in Britain, the censorship took the form of an order prohibiting bookstores from selling the book and libraries from lending it. But today, if you wanted to ban jamesjoycesulysses.com from Britain (perhaps due to a copyright claim by Joyce's notoriously litigious descendants), you'd have to look at every link that every person in the country requested, intercepting all of the nation's Internet traffic, in order to block requests to the offending site.

In China and other authoritarian nations, the national firewall isn't just a handy way to keep "destabilizing" information out of the public's hands—it's a way to spy on people, too. That's the type of practice that censorwalls would bring to the U.S.

In case you have any doubt about the throughline here: during the U.S. debate on SOPA and PIPA, the Motion Picture Association of America circulated a report written by the Information Technology and Innovation Foundation, a rightsholder-dominated think tank, that argued that the censorship measures in SOPA and PIPA would probably work, because they were the same measures already in use in China, Iran, the UAE, Armenia, Ethiopia, Saudi Arabia, Yemen, Bahrain, Burma, Syria, Turkmenistan, Uzbekistan, and Vietnam. If these measures worked there, they would work in America, too!

Not everyone minds these kind of bedfellows, of course. In early 2010, U2 frontman Bono penned a *New York Times* editorial in which he explicitly called on Western nations to produce Chinese-style national firewalls to protect fledgling artists' incomes. He wrote, "But we know from America's noble effort to stop child pornography, not to mention China's ignoble effort to suppress online dissent, that it's perfectly possible to track content" online.

This argument is what scientists call "not even wrong"—that is, it proceeds from factual misconceptions and then compounds them. First of all, America, unlike the countries discussed above, *doesn't* have a national firewall to stop child pornography. And as for China, its national firewall is a useful tool for surveilling the country, and moderately effective at preventing unsophisticated Internet users from visiting Amnesty International's website, but it is not particularly effective at keeping actual, committed dissidents offline. (China's main tool for blocking dissent is paying a vast number of shills called the "Fifty-Cent Army" to pretend to be average citizens and post floods of messages supportive of government positions on various websites.)

China is literally ruled by engineers. Eight of the nine previous politburo members were engineers. Of the twenty-one party members who sat on Politburo Standing Committees between 1992 and 2012, nineteen were trained as engineers; China as a whole has more engineers than any other country in the world, and what's more, it's the place where the majority of the world's networking equipment is manufactured. It's also a country without the rule of law, where arbitrary punishments, including indefinite detainment, are handed down for subverting the Great Firewall.

But despite all this, China has not managed to "track content." Its censorwall, at the end of the day, is no more impenetrable than the ones that have been constructed anywhere else. If China can't do this—with all those engineers, with all that equipment, with so few impediments to harsh punishment—then how does Bono expect the liberal democracies he was addressing in his *Times* editorial to manage it?

It's worth noting, too, that Bono is a famous campaigner for human rights. He has raised large sums for human-rights causes, and has taken a very public stand in favor of the fundamental rights of all people—free speech and privacy, and all the freedoms that flow from those two, such as freedom of association, a free press, and accountable, democratic governments.

There is no doubt in my mind that Bono is sincere when he advocates for those causes. I know Bono loves human rights. I just wish he'd share.

Adding censorship to the Internet means adding surveillance to the Internet. Creating Great Firewalls means creating secret, unaccountable lists of censored material that result in mass abuse, even in the most liberal of democracies. It doesn't matter if you're censoring for copyright infringement or for human-rights reports. The result is the same: a surveillance state.

Collateral damage in Moscow and the Pirate Bay

In 2006, the Swedish police raided the data center that housed the Pirate Bay, an infamous BitTorrent tracker that had made a sport of taunting the entertainment industry. The circumstances surrounding the raid were contentious: it seemed the action had been improperly ordered by a government minister who was supposed to have an arm's-length relationship with the police, at the behest of the Office of the U.S. Trade Representative.

But what was more controversial in wider Swedish society was the collateral damage of the seizure: hundreds of websites went down at the same time as the Pirate Bay, as the police enthusiastically seized a data center's worth of servers. These other servers—which hosted sites for businesses, nonprofits, and individuals—had nothing infringing on them, but the police couldn't be certain of this at the time, so they took the lot. It's like they decided that, since one store in the middle of town was carrying unlicensed products, they were going to shut down the entire shopping district while they figured things out.

This travesty was widely reported in the Swedish press, and galvanized Swedish opinion in favor of the Pirate Bay. Before the raid, Swedes were ambivalent about the site; afterward, many people decided that whatever wrong the Pirate Bay might be responsible for, it was peanuts compared to the real harms that entertainment-industry pressure had created—suborning illegal interference with the police and ordering a raid that casually destroyed the Internet presence of hundreds of local institutions.

The Pirate Bay was back online in seventy-two hours. The Pirate Party, a political party founded in the wake of the raid, now has affiliates in more than forty countries, and Sweden has elected two Pirates to the European Parliament.

A year after the raid, a wave of computer seizures by police swept across the offices of multiple newspapers and opposition groups in Russia. It seemed like a perverse tribute to the demands of the entertainment industry and the U.S. Trade Rep, which had long been agitating for more far-reaching Russian copyright legislation in advance of the country's admission to the WTO. "Our law enforcement finally realized that computers are very important tools for their opponents, and they have decided to take away these tools by doing something close to the West's agenda," Vladimir Pribylovsky, head of the Panorama Research Institute in Moscow, told the *Washington Post* in 2007. "I suppose you could say it's very clever."

In most cases, the computers were seized under the pretense that their users were running infringing versions of software like Microsoft Windows.

In the Russian city of Samara, an election-monitoring group had its offices raided just ninety minutes after one of its leaders spoke on Russian radio in support of a planned protest march. Police told her that she was under investigation for the use of unlicensed software. In almost every case, the accused argued—to no avail—that they had not been using stolen software at all.

"This is not a campaign against piracy, it's a campaign against dissent," Vitaly Yaroshevsky, a deputy editor of the *Novaya Gazeta* newspaper, told the *Post*. "It doesn't matter if we show we bought computers legally. It will change nothing."

Meanwhile, the entertainment industry cheered as Russia, in 2008, enacted its latest copyright legislation, despite (or because of) the provisions it included for warrantless search and seizure. In the version of the law that the West's entertainment lobby demanded, the burden of proof is shifted from the accuser to the accused—political dissidents are presumptively guilty of copyright infringement, and have little hope of proving otherwise.

The Problem with Cutting Off Access

REMEMBER THE DISCUSSION of "three strikes" laws, in the last section? These laws have already been stamped into the law books of France, New Zealand, and the UK. They're sort of the inverse of censorwalls—instead of keeping objectionable material off the Internet, you're keeping objectionable *people* from connecting to the web. Unfortunately, it's far from a problem-free approach.

Under Three Strikes, in theory, any ISP customer who attracts three copyright complaints is banned from accessing the Internet for a set period of time. But most Internet connections don't serve a single person: they serve a whole family, or sometimes several families. Maybe they're all pirates, or maybe only one of them is—or maybe none of them are. It could be that someone nearby is tapping into the family's WiFi, or it could be that the rightsholders bringing the complaints have just made a mistake. In 2008, the Motion Picture Association of America sent a letter to the University of Washington complaining that a certain IP address (the numerical address assigned to a computer when it connects to the Internet) was responsible for downloading a copy of *Indiana Jones and the Kingdom of the Crystal Skull*. It turned out that the IP address in question had been assigned to a laser printer—which, needless to say, wasn't downloading any movies. (Researchers at the university had "framed" the printer for the download in order to demonstrate flaws in the MPAA's enforcement mechanisms.)

In France, Three Strikes was initially established through a law called Hadopi, which was vigorously pressed by President Nikolas Sarkozy. The first version of the bill was rejected by the French Constitutional Council as a violation of the human rights of French citizens. Sarkozy's government added a fig leaf of due process to Hadopi through a "streamlined" court process for handling disputes—although those who appealed their disconnections would still be presumed guilty and required to prove their innocence.

Two years after the law's passage, a site called You Have Downloaded used the same methods as those deployed under Hadopi to trace several acts of copyright infringement to President Sarkozy's official residence. (Apparently, someone there had downloaded the Beach Boys' *Greatest Hits*.) The government disputed the accusation, even as it continued to claim that millions of French residents were inches away from expulsion from the Internet.

The difficulty—and legal expense—of proving a particular individual responsible for an act of infringement may explain why these laws have fallen out of favor. France abandoned its banning policy in 2013; up to that point, the government had followed through on only one Three Strikes case, which resulted in a mere fifteen-day ban—which was then canceled before it went into effect.

But leaked drafts of the Trans-Pacific Partnership show us that Three Strikes is still a going concern for the entertainment industry. In the UK, the Digital Economy Act allows the government to establish a Three Strikes rule without any further parliamentary action. And the latest wave of "voluntary agreements" between entertainment companies and the major ISPs around the world have shown us that what the entertainment industry can't get through law, it often gets through sheer chutzpah.

Copyright and Human Rights

WHEN VIACOM SUED Google over not doing enough to keep Viacom's copyrighted works off of YouTube, it made a claim that shocked even seasoned copyright watchers. Viacom argued that YouTube was complicit in acts of infringement because, among other reasons, it allowed its users to mark videos as "private," so that only the uploader's friends and family could see them. Private videos couldn't be checked by Viacom's copyright-enforcement bots; Viacom argued that the courts should hand over access to that material to them.

I'm pretty familiar with YouTube's privacy flag. I've used it, for example, when I wanted to show my parents in Canada and their parents in Florida and my nieces and nephews in Wales videos of my then-two-year-old daughter frolicking naked in the bathtub. I certainly couldn't attach a giant video file to an email—at the time, the privacy flag was the way to go, though a year later, file-locker services like MegaUpload had become so easy to use that I could have employed one of them to do the same thing. (Although who knows for how long, since the entertainment giants have gone to war against the file-locker services, too.)

I'm one of the millions of people who use YouTube for personal reasons. I don't have the wherewithal to set up my own miniature personal YouTube, and yet from time to time I have something personal I want to share with the people in my life. I think that everyone probably has moments like that, ones they'd like to share with a few others without putting them out there for the whole world to see.

Under Viacom's legal theory—which was supported in amicus briefs filed by organizations representing all the major studios, broadcasters, publishers, and record labels—companies should allow the giant entertainment corporations to access all of our private files to make sure we're not storing something copyrighted under cover.

This is beyond dumb. It's felony stupidity. It's like requiring everyone to open up their kids' birthday parties to enforcers from Warner Music, to ensure that no royalty-free performances of "Happy Birthday" are taking place. It's like putting mandatory webcams into every big-screen TV, to ensure they're not being used to run a bootleg cinema. It's like a law giving the big five publishers keys to every office in the land, to ensure that no one is photocopying books on the sly.

Moreover, it's a measure that no studio executive would personally tolerate. I've met plenty of people from the exec suites and the enforcement arms at the studios, and you'll not meet a more admirably paranoid bunch of IT users. If you visit Fox Studios, in Hollywood, you'll find that you have to allow your car to be searched and surrender your cameras at the gates, so as to protect the secrecy of the material within their fortress. This company, and the industry it inhabits, would never in a million years allow its own work to be subject to the kind of oversight they advocate for the rest of us.

And yet, according to Viacom's theory and its amicus backers, privacy is incompatible with copyright in the twenty-first century. They maintained that the ability of the public to communicate in private was an existential threat to creativity itself. Fortunately, the judge didn't agree—well before the case was settled, the court rejected Viacom's demand to peek beneath that privacy flag.

Regardless of the judge's decision, we should never forget what the studios and labels have told us about their beliefs about the future of copyright: that the law should forbid privacy in order to keep it from being used to hide infringement. Viacom told the court that its industry couldn't peacefully coexist with an option to keep your personal data private. Given the relentless lobbying these companies engage in, it would be foolish to think their statements about what the law should say reflect anything but what they believe the law *will* say.

A World Made of Computers

THERE WAS A time when our myriad technologies were fundamentally different. You could regulate the guts of an airplane without regulating the guts of an automobile, and no matter what you did to either of those, you wouldn't have any impact on hearing aids.

No more. Today's world is made of computers. A car is a computer you sit in. A plane is a flying Solaris minicomputer connected to a bunch of commodity SCADA controllers of the sort found everywhere from conveyor belts to nuclear power plants. Increasingly, our houses are computers that we live in.

I'm a member of the Walkman generation. Like the iPod generation that came after mine, I will almost certainly end up with a hearing aid if I live long enough to attain my dotage. When I do, my hearing aid won't be made from an analog circuit; it will be a digital computer. Why bother, after all, with expensive, balky analog circuits when you can install low-power-consumption computers running signal-processing algorithms that cancel out background noise and account for echoes in large halls—and that can let your hearing aid double as a mobile-phone headset and headphones for your laptop and TV, to boot?

You and I and most of the people we know will spend a large chunk of our lives with computers inside our bodies. We will also spend a good deal of time inside of computers, some of them moving at very high speeds.

If that's going to be the case, I want to be protected. When I put a computer in my body and put my body in a computer, I want to be sure that it is designed to take orders from its user, and to hide nothing. That's not to say that malicious parties won't find ways to hijack our future computers—security is a process, not a product—but let's not deliberately include avoidable risks in the life-support mechanisms of the information society.

Renewability: Digital Locks'
Sinister Future

THAT LINE—"SECURITY IS a process, not a product"—is a truism among security experts. As cryptographer Bruce Schneier quipped, "Anyone, from the most clueless amateur to the best cryptographer, can create an algorithm that he himself can't break." But that doesn't mean the system is secure—it just means that it's secure against people as dumb as its designer.

Which is why real security systems are designed to be patched. We know that any security system we design will have flaws in it that we can't locate, and so we need to be sure that we can update the system as new flaws are discovered and disclosed.

This applies to digital locks just as much as to any other security technology. This is why digital-lock vendors increasingly demand "renewability" in their system designs—"renewability" is a fancy way of saying that digital locks contain some mechanism to update themselves after they're shipped.

For example, if the secret keys for a specific Blu-ray model are found to be compromised, future DVDs can be shipped with "revocation messages" that prevent them from playing in that player. Likewise, future devices—recorders, amplifiers, and TVs—can be manufactured so that they no longer recognize it. From the movie studios' perspective, this is "security," since it means that new, noncompliant devices can't be made with the leaked keys—nothing else will talk to them. If you build a Blu-ray player that uses a leaked key to allow movies to be ripped and converted to play on a mobile phone (something that Blu-ray players are forbidden to do), that player won't output to your new TV or amp or projector, and new discs won't play in it.

But from the Blu-ray player's owner's perspective, this is a nightmare. It means that if someone, somewhere in the world, figures out how to steal your player's key, your TV, amp, and projector will conspire with one another to disable your device—even if you've done nothing wrong with it. There's no way to guard against this, either. To prevent this kind of loss, you'd somehow have to shop around for Blu-ray players that were secure against attacks on their secret keys. But there's no way to know whether a Blu-ray player has good or bad security. The product literature won't tell you. The Amazon reviews won't tell you. The salespeople in a store won't tell you. None of them will know—not until the keys leak, and by then it's too late. Your player is revoked, and you're up the creek.

Renewability can be even more abusive. In 2009, Amazon used a secret facility in the Kindle to delete copies of George Orwell's *1984* from the hard drives of its customers. These customers had bought their copies fair and square, but Amazon got involved in a copyright dispute over the licensing of the book (Orwell's work is in the public domain in several countries, but not in America), and its initial response was to simply reach out to the devices it had sold and arbitrarily delete the literature its customers had bought. It later thought better of the problem and reinstated the books, and promised not to do anything like that again. But Amazon won't say whether the function is still built into the device, waiting to be hijacked by someone who compromises its systems, or a court order that Amazon must comply with, or a change in management.

I worked as a bookseller for many years. In all that time, I never had to promise that I wouldn't come over to my customers' houses and take back the books I'd sold them. No court could have ordered me to do that. Nor would a crook who gained access to the store have had the ability to take away any of the books we'd sold over the years. But by building a facility for managing and enforcing copies into the Kindle,

Amazon has created a new set of vulnerabilities to legal and technical attacks that are absolutely without precedent. What's more, Amazon won't say whether the files that you load onto your Kindle yourself—books that you've bought elsewhere, personal files, other media—can also be removed by sending orders through the central server.

In 2009, Amazon announced a new feature for the Kindle. It could now use a text-to-speech program to read its customers' e-books aloud in a sort of robotic monotone. This was and is pretty cool. But some publishers objected, claiming that this was tantamount to making an unauthorized audiobook. This is a weird theory of copyright—a bit like saying that allowing users to change the font in the book is the same as making an unauthorized large-print edition. Amazon walks a thin line with the publishing establishment, though, perceived as both threat and ally, and in this case it decided to mollify the publishers.

So Amazon reached out to all the Kindles in the field and removed the text-to-speech feature from books that had already been sold.

Amazon's not the only one capable of actions like this. Apple sends out iTunes updates that remove features from proprietary iTunes music files and the players those tracks work on. For example, an early version of iTunes came with the ability to stream music from one copy of iTunes to another—you could use your home iTunes server to listen to music at work, or in your hotel room. After a negotiation with the record labels, though, Apple shipped an iTunes update that removed this feature, and restricted streaming to computers on your local network. Users didn't *have* to install the update, but if they didn't, they missed out on security patches that came out afterward. And when the next version of the operating system came out, it came with an updated iTunes that couldn't be downgraded to the more functional version.

TiVo has used renewability in similar ways. In 2006, the company pushed out an update that limited what its users could record off their TVs—if a broadcaster flagged a program as "do not record," the updated

TiVo would not permit its owner to override the request. Copyright doesn't allow broadcasters to decide which shows they'll let you record and which ones they won't, but copyright no longer matters when you've got renewable digital locks and laws prohibiting you from breaking them.

Obnoxious renewability incidents have become more common in recent years. Nintendo sells a little handheld gaming computer called the 3DS. As with Apple and the iPad, Nintendo makes additional money on the 3DS by charging game creators for the right to sell their products to 3DS owners. They enforce this gatekeeper role with a facility built into the 3DS operating system that causes it to check for special cryptographic signatures on its programs, which indicate that the Nintendo tax has been paid on the games that bear their stamp.

It's not all that technically challenging to remove this signature checking, but woe betide the child who modifies her 3DS so that she can buy or trade home-brew games. The 3DS protects Nintendo's profits by connecting to any networks it can find—even if its owner tells it not to. Then it checks to see if there's an update to its OS—even if its owner tells it not to. If it finds an update, it downloads it—even if its owner tells it not to. Then it installs the update—even if its owner tells it not to.

Once the new operating system is installed, the 3DS restarts itself and checks to see whether the old operating system was "tampered with." If it detects any tampering, it switches itself off and never switches on again. It is broken forever.

Digital locks can't work without renewability. You can't "protect" devices from their owners unless you can update them without their owners' knowledge or consent.

But renewability for digital locks means that you can't be allowed to know what's running on your computers. And that means you can't decide what's running on them. This is bad enough when we're talking about the devices we use to communicate and do our jobs. But since all computers are pretty much the same—remember "general purpose"— the endgame for renewability must be that all computers are built with this facility in mind.

Imagine what life will be like once you've got computers in your body and your body in computers. Imagine what it will mean when the person operating a car, or carrying around an implanted device, can't know or control what's running on that computer—but third parties can.

UEFI

UEFI stands for "Unified Extensible Firmware Interface." UEFI is a technical specification that computers can use when they're turned on to check the "signature" of their operating systems and make sure that it matches an approved signature from the manufacturer. This can be very useful for determining whether a computer has been compromised by a rootkit or other malicious software—infected systems will no longer be able to match the list of known-good signatures.

UEFI has additional potential, though. Depending on whether users are allowed to override its judgment, it could also be used to prevent users from running operating systems that they trust more than the signed ones. For example, you might live in Iran, and believe that the Iranian police use UEFI to ensure that only versions of Windows with a built-in wiretapping facility can run on officially imported PCs. Why not?

In 2011, a branch of the German government was caught illegally sneaking a rootkit (the *Staatstrojaner*, or "State Trojan") onto the PCs of people it suspected of crimes. Like the rootkit that the Lower Merion School District snuck onto its students' computers, the *Staatstrojaner* let its masters covertly operate their targets' cameras and mics, capture keystrokes from their keyboards, and read their files and watch their screens.

The German government had to go to some trouble to infect its suspects' PCs with the *Staatstrojaner*. Would Iran really hesitate to ensure that they could conduct *Staatstrojaner*-grade surveillance on anyone, without the inconvenience of installing a *Staatstrojaner*-type program in the first place? If they chose to, Iran could just ban the sale of computers unless UEFI was set to require surveillance-friendly operating systems out of the box. And if it can happen there, it can happen here.

A World of Control and Surveillance

THE EDWARD SNOWDEN leaks left much of the world in shock. Even the most paranoid security freaks were astounded to learn about the scope of the surveillance apparatus that had been built by the NSA, along with its allies in the "Five Eyes" countries (the UK, Canada, New Zealand, and Australia).

The nontechnical world was most shocked by the revelation that the NSA was snaffling up such unthinkable mountains of everyday communications. In some countries, the NSA is actively recording *every single* cell-phone conversation, putting millions of indisputably innocent people under surveillance without even a hint of suspicion.

But in the tech world, the real showstopper was the news that the NSA and the UK's spy agency, the GCHQ, had been spending $250 million a year on two programs of network and computer sabotage— BULLRUN, in the USA, and EDGEHILL, in the UK. Under these programs, technology companies are bribed, blackmailed, or tricked into introducing deliberate flaws into their products, so that spies can break into them and violate their users' privacy. The NSA even sabotaged U.S. government agencies, such as the National Institute for Standards and Technology (NIST), a rock-ribbed expert body that produces straightforward engineering standards to make sure that our digital infrastructure doesn't fall over. NIST was forced to recall one of its cryptographic standards after it became apparent that the NSA had infiltrated its process and deliberately weakened the standard—an act akin to deliberately ensuring that the standard for electrical wiring was faulty, so that you could start house fires in the homes of people you wanted to smoke out during armed standoffs.

The sabotage shocked so many technology experts because they understood that there was no such thing as a security flaw that could be exploited by "the good guys" alone. If you weaken the world's computer security—the security of our planes and nuclear reactors, our artificial hearts and our thermostats, and, yes, our phones and our laptops, devices that are privy to our every secret—then no amount of gains in the War on Terror will balance out the costs we'll all pay in vulnerability to crooks, creeps, spooks, thugs, perverts, voyeurs, and anyone else who independently discovers these deliberate flaws and turns them against targets of opportunity.

So where does all this tie in with the copyfight? The laws behind digital locks make it illegal to determine what your computer is doing. They make it illegal to stop your computer from doing things you don't like. And they make it illegal to tell other people about what's going on inside your computer.

As you read this, digital locks are proliferating in new and deadly ways. In 2013, the World Wide Web Consortium (W3C), a standards-making body that had always stood for the open Web, capitulated to the DRM people. After brutal arm-twisting from the entertainment industry, as well as their major partners—Google, Apple, and Microsoft—the W3C agreed that new versions of the Web would have DRM baked right into the standards.

The tech companies wanted DRM so that they could get your browser to fight you if you tried to save a Netflix or BBC iPlayer video. (Netflix and the BBC want to be sure that you watch their content only when you're connected to their services.) To make this work, they had to design a browser that was capable of disobeying its owner's orders.

Such a browser needs to be closed source, because if the code was available to users, they could change it so that the "Never save video" feature was a "Save video whenever I want" feature.

And because any information about bugs in this DRM could be used to break it, that information has to be illegal to disclose, just like the bugs in all the other DRM.

The upshot of this is that, in order to make sure we watch TV in the prescribed manner, every device with a browser-based interface is about to become a reservoir of long-lived, illegal-to-report bugs that can be exploited to attack us in every way imaginable.

When we take away the right to figure out if something bad is going on in our computers, the inevitable consequence is that bad things will happen in our computers.

Furthermore, as we observed earlier, digital locks leak. They leak unscrambled files that can be freely copied, like the movies ripped from DVDs that you can download online. They leak keys that can be used to produce unscrambled files, like the 09 F9 key. They leak tools that use those keys to unscramble files, like HandBrake.

The only way to keep files, programs, and keys out of wide circulation is to give rightsholders the legal authority to demand that files be removed without court orders, and to establish national censorwalls that surveil all Internet traffic and interdict requests for sites that rightsholders have added to blacklists. Some rightsholders make the argument that even this isn't nearly enough: in lobbying for SOPA, industry representatives argued that they also needed the right to censor DNS records, as well as a ban on tools that might defeat any of this censorship.

DNS, remember, is the service that converts human-friendly Internet addresses (like thepiratebay.se) into machine-readable numeric addresses (like 194.71.107.50). You can think of this as being akin to the way your GPS works. You tell your GPS you want to go to 1600 Pennsylvania Avenue, and it converts that to a latitude and longitude like 38.89859, -77.035971 and promptly supplies driving directions to the White House.

Playing shenanigans with DNS has lots of upsides, if you're a criminal or an oppressive government. Criminals like to hack DNS servers to redirect requests like "www.citibank.com" to lookalike webpages that they operate, so that they can get your banking details and clean you out when you unsuspectingly type in your password. Oppressive governments like to redirect gmail.com and facebook.com to their own "man-in-the-middle" servers, so that they can snoop on citizens' email and figure out whom to arrest.

Lots of people are trying to solve the DNS problem. It is real, and grave. Many Internet-security experts consider the insecurity of DNS to represent an existential threat to the Internet itself, and there are many efforts under way, like DNSSEC, to add a layer of security to the service. Your ability to vote, interact with your government, bank, get an education, and securely conduct most of the rest of your online life is dependent on the outcome of these efforts. They matter.

The objective of DNSSEC and other proposals is to detect and evade shenanigans at DNS servers. But DNS can't (and shouldn't be expected to) distinguish between the false DNS records doctored by a criminal and those instituted, in the form of redirects, by a record label or a government. Which puts it squarely in opposition to proposals like SOPA, which ban circumvention of DNS blocks.

And thus it becomes illegal to add security to the DNS system.

But it doesn't stop there. SOPA and PIPA (and related proposals) also made provision for "IP blocking." This would require ISPs to block traffic from certain known-offender IP addresses, such as 194.71.107.50, the address used by thepiratebay.se. In this scheme, ISPs would be compelled to program their routers to reject traffic from such addresses, and it would become an offense to provide tools to get around the block.

IP blocking isn't new. Like DNS blocking, it's a well-established tactic in oppressive states, a feature of Great Firewalls in China and elsewhere. The Onion Router (TOR), the proxy tool I mentioned earlier, uses clever cryptographic techniques to get around such blocks. TOR continues to be improved by a robust community of programmers, and it's the go-to tool for dissidents and everyday people around the world who want to visit sites that their governments have attempted to block. It played a critical role in the Arab Spring uprisings, and continues to be used in other fights for freedom.

TOR's job is to get you the address you request while stopping spies from knowing what you're looking at. Like DNSSEC, TOR doesn't (and can't) distinguish between "bad" censorship in China and "good" censorship in America. Establishing an American firewall, and banning work on tools to circumvent IP blocking, is effectively a ban on TOR and all its successors and competitors. It criminalizes fundamentally public-spirited work to secure the Internet for everyone in the world. And it threatens to put TOR's users in mortal danger, by criminalizing the developers who keep TOR up-to-date against attacks on it by the hired guns behind the world's firewalls.

What Copyright Means
in the Information Age

SO—WE KNOW WHAT one kind of regulation is going to look like. How do we decide what kind to implement in our world, without causing all the damage that industry policy threatens to bring down on our heads?

The fact is, we're headed inexorably toward a world made of computers and networks. That world must be one where those computers are designed to be our honest servants and the networks our honest brokers. There is still a place for copyright in this age, but when, as we've seen, every device functions by copying all the time, copyright's job can't be to "regulate copying." It was a critical mistake to assume that, as copying became integral to every corner of our world, the industrial regulation of the entertainment industry should expand to regulate everything.

There's nothing wrong with the idea of a well-developed regulatory structure for the entertainment industry and its supply chain. But copying will never, ever get harder. Hard drives in the future will not get magically less capacious, or slower, or more expensive. Networks will not get harder to use, and there won't be fewer of them, and they won't be slower. There won't be fewer people who know how to type "the hobbit iii bittorrent" into a search engine. This day, today, the day you read these words, is, from this time onward, the hardest day copying will ever face.

Your grandchildren will marvel at how hard it all was. They'll say things at Christmas dinner like "Tell me again, Grandma, about those days when you couldn't buy a hard drive the size of your fingernail in blister-packs of six for a dollar from a pegboard by the convenience-store checkout! How'd you get by without drives that could hold all the music, movies, books, photos, and paintings ever made?"

In the twenty-first century, copying isn't a problem. In the twenty-first century, copying is a *fact*. You can't and won't solve copying.

It's not impossible to regulate these things well in this world—it's just impossible to regulate these things well if all you care about is my publisher's bottom line or my royalties or the fortunes of the six movie studios that grew to global scale thanks to the last round of technological innovation.

If we're going to regulate the Internet and the computer, let's not treat them like glorified cable-TV delivery services. Let's regulate them as the building blocks of the information age.

Copyright: Fit for Purpose

LIMITING COPYRIGHT'S SCOPE frees us to design copyright rules that treat copying as a fact. These rules would regulate the industry, not the world. If we want to design a copyright to benefit publishers, we can scrap the digital-lock rules that put mere retailers in the driver's seat, giving them tools to establish and maintain choke points in the distribution of creative works; if we want to design a copyright to benefit authors, we can abolish the intermediary liability rules that make it impossible to connect with an audience unless you sign away your rights to a publisher.

Two hundred-some years ago, the United States Congress established a very clever copyright rule that was designed to protect publishers *and* authors. The rule, established by the Copyright Act of 1790, created a short period of initial copyright, fourteen years, renewable by the author for another fourteen years. This created a dynamic that I will fictionalize in a short two-act play:

ACT ONE

A publisher's office in colonial America. The PUBLISHER, wearing a green eyeshade, presides over a huge cast-iron press, standing behind a countertop surmounted by a bushel of sickly potatoes. The AUTHOR, dressed in ragged trousers, enters, carrying a large manuscript, which he thrusts upon the publisher.

AUTHOR: I have written a novel!

(Publisher takes the manuscript and peruses it.)

PUBLISHER: Not bad, my fine fellow. If you wouldst assign to me thine copyright, I will remunerate you handsomely with yonder potato.

AUTHOR: Just one potato? It took me a year!

PUBLISHER: If thou doubt mine generosity, I invite you to offer this manuscript to the scoundrels down the road. I hear they are paying authors in half carrots.

AUTHOR: Do you really need my whole copyright assigned to you? What if I just license it to you?

PUBLISHER: Ah, but this is how we protect our investment in the work. Once thou hast assigned unto me thine copyright, I can use it to sue yonder down-the-road scoundrels, who would otherwise happily put out a pirated edition.

AUTHOR: Well...

(*Publisher offers a contract in one hand, a potato in the other. The author signs the contract.*)

ACT TWO

(*The publisher's office, fourteen years later, and much enriched. Many presses clank away in the background, attended by busy, efficient laborers. The publisher has grown fat and satisfied in his dotage. The author enters, even thinner and more down-at-heels than before.*)

AUTHOR: You sent for me, sir?

PUBLISHER: Ah yes, so nice to see you again. You're looking, erm... well.

AUTHOR: I don't suppose you have any more potatoes?

PUBLISHER: Indeed, sir, I do! And all I require is that you sign this copyright renewal form, so I can post it straightaway to the Library of Congress.

AUTHOR: Yes, I see my novel in bookstores everywhere. I'd forgotten that the copyright was due to expire. That must be terrible for you, when that happens. After all, you've made such a good living from it.

PUBLISHER: If you'd just sign right here...

AUTHOR: Oh, I'll sign. But before I take pen in hand, let us discuss the matter of back-royalties and my fee for renewal.

PUBLISHER: Come now, sir. If thou wouldst not sign this paper, why, thine novel would fall into the public domain, and the scoundrels down the road would begin to print up competing editions!

AUTHOR: I don't see what that has to do with me.

PUBLISHER: Sir, wouldst you have thine novel to be published by scoundrels?

(*Author gives Publisher a significant look. Publisher squirms.*)

CURTAIN

And that, ladies and gentlemen, is one way of designing a copyright that protects investors' investment without completely screwing creators.

Term Extension Versus Samplers

THE KIND OF copyright we create determines who will profit from creativity. It also determines whose work will be poorly compensated—or even banned.

Take music sampling. Incorporating snatches of one song into another has long been an integral part of musical performance. It's the basis of styles like calypso, and it has been widely practiced in popular music, from legendary jazz solos that grab a few bars of another song to the excerpt from "La Marseillaise" that opens the Beatles' "All You Need Is Love." It's a feature in classical music, of course, where "themes" from one composer find their way into other composers' works, and it's popped up in many live performances—for example, Lukas Kmit's 2011 virtuoso moment during a violin recital in Presov, Slovakia, which was interrupted by a phone emitting the Nokia ringtone. Kmit lifted his bow from the strings, and the hall fell silent except for the ringing phone. After a moment, Kmit began to improvise on the (over)familiar song; the performance that follows is astounding. It's great music. It's also perfectly legal. Go look up the video on YouTube. You'll thank me.

Analog "sampling" is legal. Copyright systems have always recognized that it's seemly, proper, and legitimate for a few notes from one song to find their way into another. Copyright doesn't give composers or performers the right to prevent this practice.

Thirty years ago, when digital samples were first incorporated into hip-hop music, no one worried about copyright. What followed was a flowering of a new genre of music, one that has had widespread commercial, critical, cultural, and popular success. Albums like Public Enemy's *It Takes a Nation of Millions to Hold Us Back* and the Beastie Boys' *Paul's Boutique* sold millions of copies and changed the face of music. They used hundreds of samples each, and didn't "clear" (get permission for) any of them. They also popularized new musical techniques, like layering several drumbeats on top of each other to make a single hyper-beat that could never be produced with analog drums.

Since then, the screws have tightened. A combination of new legislation and new court precedents have established an ironclad industry practice of clearing every sample, no matter how minimal. As Kembrew McLeod and Peter DiCola show in their 2011 book, *Creative License: The Law and Culture of Digital Sampling*, none of the classic hip-hop albums could have been released under present-day practices. *Paul's Boutique*, one of the most commercially successful records of all time, would have lost *$19.8 million* if the label had had to pay for every sample on the disc. The world of no samples without clearance makes it financially impossible to sell an album that is anything like *Paul's Boutique*.

Most legal hiphop songs today contain no more than a single sample, or perhaps two. Anything more tends to be ruinously expensive. An entire genre of music—the sort that requires more than two samples to achieve an aesthetic effect—has no legal existence. It can be heard, but anyone who creates it is liable for titanic fines. Such artists certainly can't charge money for their work. There is a genre's worth of musicians who either don't create, or, if they do create, can't be paid for making their art. This can't be what copyright is for.

It actually gets worse than that. There *is* a large pool of music that can be sampled without permission: music in the public domain—that is, music whose copyright has expired. For most popular recordings, the original term of copyright was fifty years, which meant the companies and artists who made this music were promised fifty years' worth of exclusive rights to their productions. But thanks to record-industry lobbying (sometimes abetted by musicians), the term of copyright for sound recordings in most of the world is now ninety-five years. This extension was applied retroactively, so recordings made fifty years before had their copyrights extended just as they were about to head for the public domain.

If you're a sampler, this means that virtually every song you might want to sample is still in copyright. And now that there are only three major record labels left standing, the chances are good that the recording you need to clear is owned by one of them—and it turns out that it's transcendently hard to clear a sample unless you're signed to one of those labels, too.

So where does that get us?

- Every sample of a recording that's still in copyright must be cleared.
- Nearly every recording you've ever heard of is still in copyright.
- If it's in copyright, the labels probably own it.
- The labels won't license to you unless you sign a contract with them.

Extending the scope and the duration of copyright doesn't just criminalize a whole genre of music—it also puts the labels in charge of the only legal route open to musicians, effecting a massive wealth transfer from artists to labels.

What Works?

IF DIGITAL LOCKS and intermediary liability don't work, what does? For music, blanket licenses do. If you operate a karaoke bar and someone wants to sing Jimmy Buffett's "Why Don't We Get Drunk" at 3 a.m. (a good time for that number, I'm told), you don't have to get Buffett out of bed and dicker over whether the royalty for the performance will be ten cents or twenty-five. No, you buy a "blanket license" that covers all the copyrighted music performed on your premises, from the cover band that plays on Fridays (which is how they can take requests from the audience without having a lawyer standing by to clear the rights) to the CDs your servers play on slow afternoons.

Blanket licenses are how radio DJs are able to play music. And a closely related idea, the compulsory license, allows musicians to record cover versions of each other's songs without negotiating specific permission. In many countries, blanket licenses are collected to compensate writers for library lending, and for the use of their works in university course packs.

Here's how blanket licenses work: first, we collectively decide that the "moral right" of creators to decide who uses their work and how is less important than the "economic right" to get paid when their works are used (this is the "money talks, bullshit walks" part).

Then we find entities who would like to distribute or perform copyrighted works, and negotiate a fee structure. The money goes into a "collective licensing society."

Next we use some combination of statistical sampling methods (Nielsen families, network statistics, etc.) to compile usage statistics for the entity's pool of copyrighted works, and divide and remit the collective-licensing money based on the stats.

This is not only a proven, tried-and-true scheme for paying

rightsholders for the use of their copyrights when new technology makes controlling individual uses difficult; it also has the advantage of taking what users and intermediaries do out of the realm of the illegal and into the realm of the legal. It presents copyright as something that is easy to comply with while behaving "normally"—the way all your friends act, the way that seems natural.

ISPs around the world are desperate to stave off the legal headaches of policing their users' music downloading. If the record labels offered a collective license comparable to the ones enjoyed by nightclubs, radio stations, hairdressers, and wedding halls, ISPs could advertise that their service comes with "free downloads of all music, ever!" The ISP would then pay a per-user fee to the collective, and the users and the ISP would be legit. Just as a radio DJ can legally play music regardless of the source (recordings from other broadcasts, LPs, mixtapes from friends, and random stuff from the net are all fair game for radio play), so, too, could users use any service and any protocol they liked to download their music, knowing that rightsholders would be compensated for their activity. This means that "illicit" download services could start to focus on delivering excellent user experiences, new artist-discovery systems, and bandwidth-friendly network designs that minimize the costs borne by ISPs as a result of user downloading.

The devil, of course, is in the details. How much should each user have to pay? The sweet spot is the price at which it's cheaper to get legit than it is to skirt the law. Today, downloading services focus on hiding their activity from rightsholders and ISPs. If the focus shifted to minimizing network costs, it would raise the per-user fees that ISPs could afford to shoulder themselves while still coming out ahead. And of course, we need net neutrality to ensure that the ISPs don't try to limit which services can interact with all this licensed music, to try and monopolize their relationship with listeners or musicians.

The next question is how to divide up the money. Collective

licensing societies have a poor track record when it comes to fairly distributing the fees they collect. Historically, they've been liable to capture by the major labels, who find ways to redirect the funds that are rightly owed to smaller competitors and indies. This funny accounting is only compounded when statistical sampling is done using several different methods—surveys, network monitoring, self-reporting—since the weighting of each of these numbers can substantially alter who gets paid, and how much.

But this is the twenty-first century. If there's one signal characteristic that defines today's technology world, it's analytics. Every major tech company and ad brokerage is in the business of analyzing hard-to-measure, disparate data sources. A collecting society run with the smarts of Google and the transparency of GNU/Linux has the potential to see to it that payments are fairly dispersed.

Collective licensing is legally difficult, since it's awfully close to price-fixing, and attracts a lot of close scrutiny from antitrust regulators. Usually, some form of legal "consent decree" is required to remove the hurdles to these schemes. These consent decrees can come with conditions. I'd like to see a consent decree that stipulates that at least 50 percent of all sums dispersed to labels *must* be passed through to creators, regardless of any contractual language.

This is a complex scheme, it's true. But it has some advantages, especially when compared to the current record-industry plan, which is based on attacking fundamental human freedom in the hopes of realizing the utterly speculative piracy-free Internet that no one (apart from corporate execs and their friends in government) believes to be remotely possible.

First: a blanket-license scheme *is* possible. It's been done. A lot. It works.

Second: it pays investors. If you're running a record company, your shareholders want dividends, not pie-in-the-sky talk about the money that will pour in when "the piracy problem is licked."

Third: it pays artists. This is a policy, created by statute. The statute can be designed to protect creators' income.

Fourth: it encourages investor competition. A world with three major labels is stupidly anticompetitive. A level playing field, with equal access to distribution schemes for all, makes it easy for new kinds of businesses to emerge.

Fifth: it encourages intermediary innovation. Any company that wants to produce a music service today must first negotiate a deal with the big three, a Herculean task that is often more expensive and difficult than building the service itself. It's been more than a decade since a couple of teenagers invented Napster, and practically nothing in the market today matches it for ease of use and experience. Better music services will bring in more listeners and more license fees, and pay more artists and more investors.

Copyright's Not Dead

BLANKET LICENSES ARE just one example of how we can craft new regulations for the entertainment industry, regulations that value creation, investment, and innovation without criminalizing fans or attacking the Internet. The Internet era is not—and should not be—silent on the question of how we ensure that creators and investors get a chance to make money. That's all copyright ever really wanted an answer to.

The problem is that entertainment companies have treated the increased ease of copying as a signal that copyright should be expanded to cover more people and more activities, far outside the entertainment industry. What they should have done is soldiered on regulating themselves, without trying to regulate the whole world at the same time.

Think like a dandelion

When my daughter was born, I became keenly aware of how much stock we mammals put into the copies we make of ourselves (yes, a child isn't a "copy" exactly, but go with it for a moment). Mammalian reproduction is a major event, especially for us primates, and we want to be sure that every "copy" we make grows up healthy, strong, and successful.

But there are other life forms for whom copying is a lot more casual. Dandelions produce two thousand seeds every spring, and when a good stiff breeze comes around, those seeds are blown into the air, going every which way. The dandelion's strategy is to maximize the number of blind chances it has for continuing its genetic line—not

to carefully plot every germination. It works: every summer, every crack in every sidewalk has a dandelion growing out of it.

When copying gets easier, it behooves us to adopt strategies that thrive on cheap copying. There are lots of people out there who might want to buy your work or compensate you in some other way—the more places your work can find itself, the greater the likelihood that it will find one of those would-be customers in some unsuspected crack in the metaphorical pavement.

My favorite example: I once got an email from someone who said, "I got a spam message, and pasted into the bottom of it (to fool my spam filter) was a block of really interesting text. I read it and decided I wanted to find out who wrote it, so I plugged a couple sentences into a search engine and found you. Long story short, I edit a magazine, and would like to pay you a dollar a word to reprint the essay. Is that all right with you?"

The copies that others make of my work cost me nothing, and present the possibility that I'll get something. So I try to think like a dandelion.

Every Pirate Wants to Be an Admiral

IT'S NOT AS though this is the first time we've had to rethink what copyright is, what it should do, and whom it should serve. The activities that copyright regulates—copying, transmission, display, performance—are technological activities, so when technology changes, it's usually the case that copyright has to change, too. And it's rarely pretty.

When piano rolls were invented, the composers, whose income came from sheet music, were aghast. They couldn't believe that player-piano companies had the audacity to record and sell performances of their work. They tried—unsuccessfully—to have such recordings classified as copyright violations.

Then (thanks in part to the institution of a compulsory license) the piano-roll pirates and their compatriots in the wax-cylinder business got legit, and became the record industry.

Then the radio came along, and broadcasters had the audacity to argue that they should be able to play records over the air. The record industry was furious, and tried (unsuccessfully) to block radio broadcasts that lacked explicit permission from recording artists. Their argument was "When we used technology to appropriate and further commercialize the works of composers, that was progress. When these upstart broadcasters do it to our records, that's piracy."

A few decades later, with the dust settled around radio transmission, along came cable TV, which appropriated broadcasts sent over the air and retransmitted them over cables. The broadcasters argued (unsuccessfully) that this was a form of piracy, and that the law should put an immediate halt to it. Their argument? The familiar one: "When we did it, it was progress. When they do it to us, that's piracy."

Then came the VCR, which instigated a landmark lawsuit by the cable operators and the studios, a legal battle that was waged for eight years, finishing up in the 1984 Supreme Court "Betamax" ruling. You can look up the briefs if you'd like, but fundamentally, they went like this: "When we took the broadcasts without permission, that was progress. Now that someone's recording our cable signals without permission, that's piracy."

Sony won, and fifteen years later it was one of the first companies to get in line to sue Internet companies that were making it easier to copy music and videos online.

I have a name for the principle at work here: "Every pirate wants to be an admiral."

It's Different This Time

WHEN YOU ASK entertainment execs about the historical pattern of yesterday's infringers going legit and becoming the official entertainment industry, they bridle. Yes, yes, Hollywood was founded in California so that the film companies of the day could rip off Edison's patents with impunity, but that was *different*. The Internet makes copying so *easy*, and it is so *global*, that this time the technology must be tamed or rejected, not accommodated.

I think they're half-right. This time, the technology *is* different. But not in the way they think. We are remaking the world and everything we do in it. In the past, a regulation applied to VCRs would impact a few other industries or activities (making it hard, say, to record a home movie), but it wouldn't have changed *everything*. You could regulate the VCR or the radio or the record player without regulating the automobile, the hearing aid, and voting machines along with them.

That's not true anymore. The stakes for getting copyright right have never been higher. There has never been a fight over entertainment-related technology where the consequences *for everyone outside the entertainment industry* were potentially more disastrous than they are now.

This is the fundamental tactical error on the part of the entertainment companies. Every day that goes by creates more people for whom the Internet is a key part of life. Meanwhile, the entertainment companies have told the world that unless they get to regulate the Internet, they will die. "It's us or the Internet," they say. The danger is that if they keep this up, they'll be right.

All Revolutions Are Bloody

THE INFORMATION REVOLUTION isn't bloodless. The record player and the radio made some performers rich and famous, but they also shrank the market for some types of live performance—vaudeville, for one—and doomed otherwise successful performers to penury. No copyright system and no technology has ever provided an income for even a midsize slice of all those artists who wished to earn a living from their creations. What technology does, mostly, is open doors for new artists who had been shut out of the old system—which sometimes means closing doors for the artists who had thrived.

Consider what happened when movies came along. Before movies, the only way to perform a story, while letting your audience see your performers, was to put it on a stage (or, I suppose, in a puppet theater). There were stories that wouldn't fit on the stage at all, of course, and stories that could be shoehorned onto a stage but were not exactly comfortable there. And there were also many stories that belonged onstage, and thrived there.

When film came along, stories that couldn't be performed on the stage suddenly had another place where they might be shown. This brought with it the possibility of creative success (commercial and artistic) for more creators, but it also increased the competition for audiences. This happened again when sound came to films, and again with the advent of television, and again, more recently, with the advent of Internet video. Stories that were better suited to the new medium migrated to it, and found a new chance at expression. But also: competition for viewers increased, and it became that much harder to get an audience through any particular set of doors. But the potential audience grew, too—there were people for whom stage plays didn't resonate, but who were enthralled by the moving picture.

The Internet and the PC have created entirely new media. These can produce new aesthetic effects, with new techniques that are native to new creators and new creations. And for each one, there will be different regulatory demands.

Cathedrals Versus the Protestant Reformation

REMEMBER WHERE WE started, with the battles over the Bible? The printing press made it possible for many more people to read and interpret scripture; it also sped up the fragmentation of the Church, allowing dozens of small-c churches to compete with the Church for authority, and for the resources and attention of worshippers.

Prior to that fragmentation, the Church had enjoyed access to almost unimaginable resources. The most visible expressions of that wealth and power are the grand cathedrals, built over the course of several lifetimes—private buildings whose construction consumed the entire working lives of thousands of men, generation after generation. It's almost unimaginable by today's standards.

To stand in one of these structures is to be overawed. When the age of cathedral-building drew to a close, humanity lost something. Even I—a committed atheist—can see that.

But humanity gained something, too. The Reformation ushered in an era of freedom of thought and conscience, and, what's more, it enmeshed religion more thoroughly in the lives of congregants. The churches that propagated across Europe were humble "wee kirks," rough-hewn things that were as far from the grandeur of a cathedral as you could get. But that was just how the reformers wanted it. For them, the personal religious experience was more important than the grand one. The intimacy and scale of Reform kirks offered a new kind of fulfillment. Cathedrals had their virtues, but they weren't as important as that personal experience.

Three-Hundred-Million-Dollar Movies

A FAMILIAR REFRAIN from the entertainment industry is that Internet-native video isn't worthy because it lacks the expense, the grandeur, and the production values of the three-hundred-million-dollar movies that fill our cinemas every summer. The studio heads say that without the analog dollars they depend on from theatergoers, the digital dimes won't add up to enough money to produce these media cathedrals that require the labor of thousands of talented people and the infusion of so much capital.

They may be right. As might the record execs, publishing execs, gaming execs, and other media representatives who claim that their particular, expensive, high-coordination-cost, high-production-value media require a very different kind of Internet ecosystem if they are to continue to exist. The existence of a much cheaper and much less regulated medium that competes for attention, funds, advertisers, and resources necessarily threatens the expensive, winner-takes-all incumbents of the late twentieth century.

If they are right, I will miss their works. If I have to take my grandchildren to a museum to show them the cathedrals of the last days of the pre-Internet entertainment industry, I'll feel some sadness for what they're missing.

But not too much sadness.

The Internet is making it possible for more people to write more stories, make more movies, and record more songs than ever before. It is making it possible to have deeply personal, moving, and entertaining experiences in new ways. If I have to choose between twenty hours' worth of blockbusters every summer and sixty hours of "personal" video every *second* on YouTube, I'll choose the latter.

I don't think we have to choose, incidentally. I think that old media will continue to find a home in the new world. After all, 2011 was the year that a *silent film* (*The Artist*) swept the Oscars. Despite the waxing fortunes of recorded performances, live music and live theater continue to thrive—indeed, taken as a whole, these industries are bigger than they have ever been.

The entertainment industry has a long history of characterizing its profit-maximization strategies as do-or-die existential crises. If we can't control the printing press/the record player/the radio/cable/VCRs/the Internet, they say, we will die. The reality is more like "If we can't control these things, we'll have to invent some new ways of making money, and make less from the old ways." It's happened many times before.

What Is Copyright For?

THE PURPOSE OF copyright shouldn't be to ensure that whoever got lucky with last year's business model gets to stay on top forever. Live music is great, but what a rotten thing it would have been if the winners of the live-music lottery in 1908 had been allowed to strangle recorded music to protect their turf.

I think we can tell a good copyright system from a bad one by what kind of work gets made under its rules. A bad copyright system has fewer creators making fewer types of work, enjoyed by fewer people. A good copyright system is one that enables the largest diversity of creators making the largest diversity of works to please the largest diversity of audiences. It's fine for copyright to try to secure some income for the tiny percentage of creators who'll earn a living from their work. But while it's at it, it shouldn't get in the way of all the people who are making art because they want to express themselves.

Money can't be the sole determinant of whether copyright is working. Imagine a regulatory system that gave only one company the right to make movies—a company that then made only one movie every year, which, by dint of its rarity, pulled in twenty billion dollars worldwide. Who, apart from the lucky film company's employees, would prefer that to a regulatory system that let *lots* of movies be made by *lots* of people?

When movies were invented, there was one man who claimed the right to authorize the production of films—Thomas Edison, who held key film-related patents. Edison tightly controlled how many movies could be made each year, and what subjects these movies could address. The filmmakers of the day hated this, and they fled west to California to escape the long arm of Edison's legal enforcers in New Jersey. Filmmakers with names like William Fox (Fox Studios), Adolph Zukor (Famous Players), and Carl Laemmle (Universal Studios) founded the great studios of the day because they believed that their right to expression trumped Edison's proprietary rights to the technology they wanted to use.

The big six are rightly proud of their maverick history. They agree, it seems, that "one" isn't the right answer to "How many companies should be in charge of the movie business?" Apparently, though, they think the right answer is "six." I think the right answer is "millions." Yes, we disagree on the correct number, but we are in firm accord on the idea that there is a number that can be too low, regardless of proprietary rights, profits, or other considerations. With too few firms choosing which movies can be made, lots of wonderful films won't be produced, and we'll all be poorer for that—spiritually and monetarily.

I think that big movies and little movies, and big books and little books, and big music and little music, can all coexist. I think we can have a copyright that regulates these industries with that in mind.

But the entertainment industries keep saying that their demands are the existential minimum. Give us a kill switch for the Internet, they say. Give us the power to surveil and censor, the power to control all your devices, the right to remake general-purpose networks and devices as tools of control and spying, or we will die.

If we have to choose between that vision of copyright and a world where more people can create, more audiences can be served, where our devices are our honest servants and don't betray us, where our networks are not designed for censorship and surveillance, then I choose the latter. I hope you would, too.

HISTORICALLY, THE ARTS have been on the side of free speech and privacy. Creators and their investors have long fought against book-banning, censorship, surveillance, and restrictions on creative and intellectual expression. Laws that compromise these freedoms threaten art itself. They harm us, demean us, and weaken us. Some of the proudest moments in the history of the arts have come when audiences and artists, as a community, have pulled together to defend free expression, even for works they didn't like very much—even for works they detested. Evelyn Beatrice Hall said, "I disapprove of what you say, but I will defend to the death your right to say it." In the arts, we say, "I hate what you've created, but I will defend to the death your right to create it."

Whatever challenges the Internet raises for our commercial fortunes, we must not sacrifice our free speech, freedom of expression, and freedom of association in order to improve our own bottom lines.

Not least because we'll fail. The last fifteen years saw a long list of tighter controls that yielded up only more virulent forms of copying and less control over artistic work. Napster had centralized log-in servers that could have been used to collect fees for music downloading. When those central servers were revealed as a legal vulnerability—because a court could order them taken down—new protocols like Gnutella emerged that had no central servers. When the companies working on these protocols were sued, new tools sprang up that were harder to monitor and interdict, like BitTorrent (only to be replaced by "trackerless torrents," which are even more difficult to stop). At each stage, the fight against copying has produced copying tools that are more fraught, harder to make money from, harder to monitor—harder for the entertainment business to use. Anti-copying efforts are breeding the digital equivalents of antibiotic-resistant bacteria—turning relatively benign, business-friendly technologies into systems that are designed to be as hard as possible to cooperate with.

But more important, content-blocking and surveillance are the province of book burners and censors, not creators and publishers. We have fought for generations for the freedom of conscience necessary to have a robust intellectual and creative sphere. Our forebears risked jail, violence, even death for these freedoms. We owe it to them— and to our children—to pledge ourselves anew to these values in the era of the Internet.

Edward Snowden taught us that the Internet could be harnessed and turned into an intrusive and terrifying surveillance mechanism. And since the Internet is likely to be a fixture in our lives and the lives of our children, we all have a duty to stop arguing about whether the Internet is good or bad for us and our particular corner of the world—a duty to figure out how to make the Internet into a force for helping people work and live together, with the privacy, self-determination, and freedom from interference and control that are the hallmarks of a just society.

It's not enough for creators and their industry to love free speech. We have to learn to share it, too.

What Does the Future Hold?

HERE'S A LITTLE secret: science-fiction writers are *terrible* at predicting the future. But that's okay. Everyone is terrible at predicting the future. Every significant fact about the future is unguessably weird. Only the trivial is subject to extrapolation.

I have no predictions for what the future holds. But I have hopes and I have fears, and they're both anchored by the same observation: that computers and networks make it easier for us to work as groups.

That sounds trivial, I know, but working efficiently in groups is the oldest dream our species has. When some distant ancestor of ours in the savanna hit on the strategy, it benefited everyone—some monkeys could forage for fruit, others could watch for predators, and a third group could watch the kids, and everyone got more done.

This demanded that the monkeys spend a certain amount of time making sure that their comrades were doing what they were supposed to be doing—a certain amount of doubling-back to make sure that there really was someone standing lookout, a certain amount of coming down from the lookout perch only to discover that everyone was watching the kids and no one was gathering fruit.

But it was worth the wastage and inefficiencies. That's because working with others makes us superhuman (or supersimian, as the case may be). I mean that literally. There is only so much any one of us can do on his own, but when we work with others, we can transcend that limit.

All of human history since then has been a struggle to figure out how to coordinate larger groups with lower coordination costs—fewer hours in meetings, fewer duplicated efforts, fewer moments when you discover that for the past six hours you've been pulling north and your buddy has been pulling south.

Computers and networks have come closer to solving our coordination problems than nearly any technology before them. They outperform the chalkboard, the org chart, the telephone—everything I can think of except for language, the original coordinative technology ("You go this way, I'll go that way").

When I was a young activist in the 1980s, approximately 98 percent of my time was spent stuffing envelopes and writing addresses on them. The remaining 2 percent was the time I could use to figure out what to put in the envelopes.

Today, we get all that coordinative effort gratis. (It's called the "cc" line in your email client.) It's so easy that we don't even notice it. In my lifetime, the technology to coordinate people for positive change has gone from the rolling log to the supercharged V-8, and it shows no sign of slowing.

Of course, it's not just activists who get this increase in coordinative power. And not all activists are created equal—the power to coordinate accrues equally to the NAACP as to the KKK.

But cheap coordination is, on balance, good for us. Hierarchy, form-filling, meetings, memos—these are not the source of our species's glory. They are the cost we must endure to attain glory. Anything that minimizes the drag on our collective efforts is to be celebrated.

Nevertheless, I have fears. I fear that the people who have figured out how to attain superhuman powers through the application of meetings, memos, and bureaucracies will aggressively acquire technology and use it to disrupt any upstart who threatens to accomplish the same effect with less drudgery.

After all, when you're in the business of solving problems for a living, your job becomes making sure that the problems never go away, or you'll be out of business.

So I fear that totalitarians, spooks, bullies, and plutocrats will use information technology to spy on us, to sow discord among us—and, at the extreme, to kidnap and torture and murder us. That's what happened in the Arab Spring, when repressive governments under threat of a popular uprising realized that they could mine Facebook and Gmail to figure out who knew whom in the activist world, making it easier to round them all up when the time came.

That's my fear: technology magnifying the power of the powerful—not just governments, but the record companies, the movie studios, and the online intermediaries who are increasingly shaping the creative sphere themselves—to the disadvantage of everyone else. It's an old fear. It's the fear that gave Orwell the impetus to write 1984. We are social and sociable creatures, and our networked devices keep our friends and loved ones at our fingertips all the time. Those devices accumulate a record of all our thoughts, deeds, journeys, and relationships. If they are designed to betray us, or can be subverted to do so, they make the dreams of the Stasi seem like amateurish crayon drawings. In a world of treacherous devices and networks, you don't have to choose: you can have a future that's made up of equal parts Orwell, Kafka, and Huxley.

But I also have hope. I have hope for people who want a world with more self-determination, more freedom, more accomplishment, and less crushing boredom and control. I have hope because those without power have always lacked the ability to cheaply organize themselves. Progressive politics have always been a mire of internecine struggle and endless talking-shops, because when you lack a hierarchy and an organization, you have to invent the means of getting stuff done every time you want to do something.

I think that giving people who yearn for freedom the power to organize is fundamentally different from increasing the organizational power of those who love control (usually because they're the ones in control).

When the powerless get power, it is a difference in kind. When the powerful get power, it is a difference in degree. Power where there was none is different from a little more power where it was already concentrated.

I don't think the conclusion is foreordained. On bad days, I'm petrified of the extent to which a despot could use technology to perfectly spy, to perfectly coordinate an army of thugs.

But even on those bad days, I believe that the only answer to this fear is to seize the means of information and ensure that technology's benefits are distributed to everyone, not just the powerful. A refusal to engage with (or protect) technology doesn't mean that the bad guys won't get it—just that the good guys will end up unarmed in the fights that are to come.

Edward Snowden, our only credible authority on the capabilities of the world's spy agencies, tells us that cryptography works. Good, secure networking technology allows everyday people the power to communicate with one another with such a high degree of security that even the most powerful, most adept surveillance agencies in the world can't spy on them. Anything that can keep out the spies can also keep out crooks, voyeurs, and other creeps.

In other words, for the first time in human history, average people can coordinate with each other to get stuff done without worrying about anyone else horning in on their communications channels to disrupt or destroy them. This is a technology dividend beyond all price, a treasure of historically unprecedented value.

But we get this dividend only if the infrastructure is free and fair. We get it only if we give up on the Hollywood-versus-Google narrative, and stop letting our creative output be hijacked in the service of censorship, surveillance, and control.

I'm not anti-regulation. But we need to decide what kind of regulation we want. The Internet can have rules that encourage centralization—rules permitting network discrimination, rules protecting digital rights management, rules providing for easy takedown—and, with them, rent-seeking, abusive sharecropping, spying, and censorship. Or it can have rules that promote an open, pluralistic, networked public space where anyone can communicate; rules that encourage disclosure of security vulnerabilities; rules that encourage competition by allowing interoperable products and technologies.

It has never been cheaper to make art, and it's never been cheaper to reach an audience. There's never been such broad and deep access to the creative output of our artistic forebears. As always, almost everything anyone does to get a living out of the arts won't work—the Internet doesn't change that. What it *does* change is how many ways there are to make things, and to get them into other people's hands and minds. It changes how many people can participate in culture and satisfy their creative urges.

This book is not meant to tell you what business model to pursue in your artistic endeavor. That's a moving target, and nothing would date this volume faster than overspecifying how you could use this tool or that one to maximize your return.

The world is full of people who are offering you ways to try and turn your creativity into dollars. What I hope I've given you is a tool kit for *evaluating* all those offers. Here is that tool kit, boiled down to three points:

1. If you're a publisher, don't let your retailers usurp your relationship with your customers by using DRM.
2. If you're a creator, don't let your publishers use your copyright as an excuse for rules that let it corner the market on delivering your art to your audience.
3. And no matter who you are, remember that this Internet thing is bigger than the arts, bigger than the entertainment business—it's the nervous system of the twenty-first century, and, depending on how we use it, it can set us free, or it can enslave us.

I MENTIONED EARLIER in this book that every few years, the U.S. Copyright Office holds an obscure set of hearings called the Section 1201 Rulemaking Proceeding. The purpose of these hearings is to determine whether exemptions should be granted to Section 1201 of the Digital Millennium Copyright Act.

DMCA 1201 is a bizarre rule, even by the standards of copyright law—itself a reservoir of deep weirdness. Dr. Pam Samuelson, a professor at UC Berkeley and one of America's leading copyright scholars, recently told me that she believes that *no one in America understands all of American copyright law*. At more than one thousand pages, it is ungraspably complex.

1201 is the part of the DMCA that makes it illegal to remove digital locks. Violations of 1201 are felonies, punishable by up to five years in prison and $500,000 fines for a *first offense*. The statute makes it a crime to remove a lock even if you own the copyrighted works behind it. Even helping someone else remove a lock is a crime: just publishing information about programmer errors in a digital lock can trigger the statute, since if you know about an error in the program, you might be able to exploit that error and bypass the program's restrictions.

This rule was originally supposed to apply to things like DVD players. Movie studios could embed "region codes" in DVDs, making them compatible with only those DVD players that checked for the code and refused to play discs from out of region. That way, discs that were offered cheaply in one part of the world couldn't be played in places where they were more expensive. Nothing in copyright law prohibits bringing a DVD from a poor country to a rich country, but because breaking the lock on a DVD player is illegal, and because you need to break the lock in order to watch an out-of-region disc, 1201 was meant to prevent that kind of migration from happening.

It didn't work, of course. Fly-by-night companies broke their license agreements with the studios and manufactured "region-free" DVD players, and soon the respectable manufacturers like Sony started to make their own players easy to convert to region-free. The studios rattled their sabers and then mostly gave up on the idea of region enforcement.

But the law lives on, and the studios are now much more aggro about using it to go after anyone who "jailbreaks" new formats like Blu-ray than they were about cracking down on region-free DVD players. More importantly, though, *other* industries have seized on 1201 as the answer to questions that, until this decade, were too weird to even contemplate asking.

To see where DMCA 1201 has gotten to in the seventeen years since its passage, one need only look at the docket in the 2015 Copyright Office 1201 Rulemaking Proceeding, where various researchers, scholars, and public interest groups have lined up to petition for the right to jailbreak devices that have acquired 1201 coverage.

First up: tractors. Farmers are asking the Copyright Office whether they are allowed to jailbreak their John Deere tractors. No, seriously. John Deere tractors are, fundamentally, giant, weird-looking computers, filled with complex software, and John Deere sees them, fundamentally, as "platforms" for offering "services." They want to be able to decide who offers services on their platform, so they use digital locks to stop farmers from changing the software on their devices.

The John Deere saga began when a farmer called up John Deere tech support because his tractor's tire-pressure sensor was erroneously reporting that he had a flat, and his tractor was refusing to move until the problem was resolved. ("I can't let you do that, Farmer Dave.") Deere agreed that the sensor was faulty, and told the farmer to expect a replacement part in a day or two. When the farmer asked whether he could just disable the malfunctioning sensor, he was told no, absolutely not. Deere won't allow you to change your tractor's software, and they won't let anyone else offer a product that allows you to do this, either.

Now, Deere doesn't make any money from spuriously immobilizing its customers' farm equipment. But Deere makes a *fortune* by exercising total control over the tractor/platform. The torque sensors in their tractors' wheels are able to create centimeter-accurate soil-density maps of farmers' fields, but Deere won't allow farmers to see this data. Instead, they extract it themselves and sell it to seed companies like Monsanto, who sell it back to farmers as part of a package deal that requires that the farmers use their seeds.

The data on John Deere tractors isn't copyrighted, or even copyrightable—facts aren't entitled to copyright protection. It is generated by the farmers, using equipment that they own. It's their real property. But their property rights are trampled by the metaphorical "intellectual property" rights of John Deere.

When the farmers asked the Copyright Office for an exemption allowing them to alter the software on their tractors, though, a host of industries weighed in to take Deere's side. The automakers were particularly adamant.

You should go google GM's filing in the docket. They *really* hate the idea of letting you change the software in your car (which is, of course, also a computer). GM doesn't gather soil-density data, but they *do* use software locks to control which mechanics are allowed to service GM cars. It's a felony to jailbreak a car and read GM's diagnostics info without using the official GM diagnostic tool. GM will give mechanics access to that tool only if the mechanics sign a contract promising to buy GM parts. In the GM-verse, you literally can't own a car's software—you can only license it, subject to the terms enforced by the locks on it. They weren't kidding when they said that it's "not your father's Oldsmobile."

The Copyright Office is due to render a verdict on these exemptions by Christmas 2015—perhaps by the time you read this. But even if they grant exceptions to farmers and mechanics, it's cold comfort: the Copyright Office can legalize *using* a lock-breaking tool, but it will remain illegal to *make, possess, sell, or give away* such a tool. Then there's the fact that all 1201 exceptions need to be pleaded for anew every three years, so any relief an exception *could* afford will expire in 2019.

If you're worried only about how 1201 hurts your wallet or interferes with your weekend car-tinkering, though, you're not thinking big enough.

Computer security is transcendentally hard. Every day, we learn about a new, awful bug in the systems we rely upon. That's because there's only one experimental methodology for validating our security work: peer review. To paraphrase Bruce Schneier, it's easy to design a security system that you can't think of any flaws in. Until you tell other people how it works, all you know is that it'll work on people dumber than you. As in all forms of engineering and science, adversarial peer review—where your friends tell you about the mistakes you've made, and your enemies excoriate you for stupidly making them—is how we validate and further our knowledge.

Security researchers are the first collateral victims of 1201. Their daily activities—uncovering flaws in products and publishing them, after first informing the manufacturers and giving them an ultimatum ("We're publishing this bug in two months, fix it by then")—is short-circuited by a system in which manufacturers can threaten researchers with prison sentences and fines for publishing bug reports.

As we've learned, this doesn't stop bugs from being discovered; foreign spies, hackers, criminals, voyeurs, and identity thieves are all hard at work discovering and weaponizing them, DMCA or no. It just means that it takes longer before you and I are able to take measures to protect ourselves. Preventing bug disclosure by independent researchers gives the bad guys more time to attack all of us—and depending on the bad guy, the attacks can be awfully undirected. Extortionists can monetize bugs in networked home-security cameras by scanning all the ones they can find, using software to locate likely instances of nudity, then threatening those users with exposure. It doesn't matter if you have "nothing to hide"; you still probably don't want naked pictures of yourself plastered all over the net.

Which brings me back to GM. In the months following GM's filings in the 1201 hearing, researchers published a series of terrifying reports on vulnerabilities in networked cars. Chrysler had to recall 1.4 million vehicles because they could be compromised wirelessly via a bug that left the attacker in control of their brakes and steering. Then independent researcher Samy Kamkar showed how he built a $32 device called the RollJam that lets him open virtually any keyless entry system. Another researcher discovered that he could "cut the brakes" of Internet-connected cars that used a small wireless device supplied by insurance companies to monitor your driving.

Those are the bugs we know about, because the researchers were willing to talk about them, despite the risks presented by 1201. The Copyright Office docket, though, is full of researchers who *don't* want to talk about what they've found.

Right now, chances are that you or someone you love is walking around with a networked medical implant: an insulin pump, a defibrillator, a pacemaker, a cochlear implant. The companies that make those devices use DRM because they want to ensure that only they get to sell doctors the software used to read their products' telemetry and diagnostic info. According to one researcher in the 1201 docket, Jay Radcliffe, *40 percent* of the code in these devices has never been audited. Radcliffe has Type 1 diabetes, but does not wear an insulin pump, despite the fact that it would doubtless add years to his life. He's looked at the code in these devices—which are capable, in theory, of delivering a lethal insulin overdose—and concluded that they were unfit for service. But he won't tell you why.

Another researcher's 1201 filing declined to even state which area s/he worked in, because even mentioning it was likely to attract a legal threat. Remember, these are the good guys who tell us when the stuff we depend on is unsafe and not to be trusted.

Copyright law is weird, but it shouldn't be this weird. I make my living from the arts, mostly by writing novels, and I am all about ensuring that creators get their due.

The 2015 1201 hearings are an important turning point in the long-running, Internet-driven copyright wars. The dumb stuff that big entertainment companies cooked up in the waning years of the last century to maximize their profits has made them rich beyond the dreams of avarice, while simultaneously making it harder for creators to get a fair share of that pie. Now their so-called "anti-piracy" laws are presenting a uniquely twenty-first-century hazard, extending an invitation to every company and every industry to remake itself as a "copyright owner" and their customers as potential pirates. This legal fiction threatens to put us all in terrible, immediate, urgent physical danger.

We've just lived through more than fifteen years of copyright wars, kicked off by the Napster panic of '99. During that time, the debate has increasingly been framed as "Internet companies versus entertainment companies." In this book, I've tried to show that neither the Internet companies nor the entertainment companies are artists' friends. Both are staffed by a mix of people who care and don't care about creators getting a fair deal, but in both cases, their *overall corporate behavior* is dictated by making themselves as profitable as possible. If that profit comes by taking a bigger slice of the Internet pie through attacks on their industrial rivals, so be it. But neither side is averse to fattening its coffers by taking advantage of us, the creators who make the "content" [shudder] that keeps the net going.

All too often, creators are convinced to take the side of the "creative industries" in fighting against "piracy." We let Hollywood use us as a club against Silicon Valley. That just means that it's Hollywood, not Silicon Valley, that gets to take advantage of us. Meanwhile, Hollywood's conception of information feudalism—where you can't ever own anything, only license it on terms set forth in an abusive, farcical "license agreement"—has merged with Silicon Valley's, and colonized everything from light bulbs to insulin pumps, tractors to coffee makers.

We did that. Us. Creators. We own that. We created the moral reality that treating ideas as eternal corporate property was the natural course of things. We did it because we thought it would stop "thieves" (that is, fans) from "ripping us off" (that is, reading, listening, watching, and playing our work in unauthorized ways).

Some will say that the rise of information feudalism was an unforeseeable consequence of artists' thrashing at the start of the Internet era. But plenty of people foresaw it then, and they're fighting it now.

A new generation of artists is carving out a new path online. They're novelists who build an audience on YouTube and leverage it into six-figure publishing deals. They're musicians who supplement their gigging by teaching students over video links. They're game designers who Kickstart games that no publisher would touch.

Most of them fail. Most artists have always failed. But the lesson these indies offer is that we don't *need* a Big Content–compliant Internet for artists to thrive—and that your best way of getting more out of a big studio or publisher is to have the credible possibility of walking out the door and doing it on your own.

I wrote this book to explain how artists can make a living from the net, but, more importantly, to explain how we can work to prevent this crazy, awful stuff from being done in our names. The fight continues. Help us win it.

Cory Doctorow
Burbank, California
August 2015

ACKNOWLEDGMENTS

Thanks to Nat Torkington, Chris DiBona, Alice Taylor, Ang Cui, Martha Lane Fox, Wendy Seltzer, Brewster Kahle, Russ Galen, Eric Faden, Matt McLernon, and Lisa Gold.

ABOUT THE AUTHOR

Cory Doctorow is a science-fiction novelist, blogger, and technology activist. He is the coeditor of the popular weblog *Boing Boing*, and a contributor to the *Guardian*, the *New York Times*, *Publishers Weekly*, *Wired*, and many other newspapers, magazines, and websites. He is a special consultant to the Electronic Frontier Foundation, a nonprofit civil liberties group that defends freedom in technology law, policy, standards, and treaties. He holds an honorary doctorate in computer science from the Open University (UK), where he is a visiting professor; in 2007, he served as the Fulbright chair at the University of Southern California's Center on Public Diplomacy. His young-adult novels include *Homeland*, *Pirate Cinema*, and *Little Brother*; his novels for adults include *Rapture of the Nerds* and *Makers*.

craphound.com
boingboing.net